Genetic Engineering

MANIPULATING LIFE

Where Does It Stop?

Cloning
Test-Tube Babies
Abortion
Surrogate Mothers

By
Duane T. Gish, Ph.D.
and
Clifford Wilson, Ph.D.

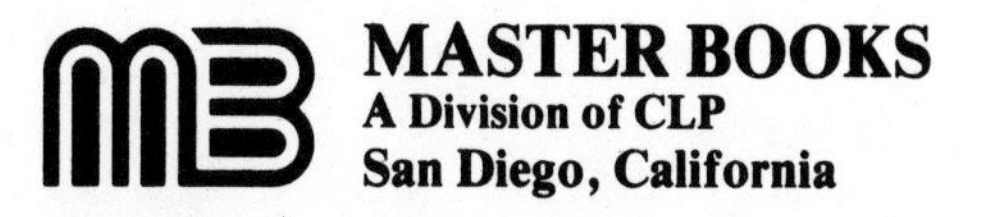

MASTER BOOKS
A Division of CLP
San Diego, California

MANIPULATING LIFE: Where Does It Stop?

MASTER BOOKS, A Division of CLP
P. O. Box 15666
San Diego, California 92115

Library of Congress Catalog Card Number 81-65488
ISBN 0-89051-071-7

Cataloging in Publication Data

Gish, Duane Tolbert, 1921-
Manipulating life: where does it stop? [text] / by Duane T. Gish and Clifford A. Wilson.

1. Bioethics. 2. Cloning—Religious aspects. 3. Abortion—Religious aspects. I. Title. II. Wilson, Clifford A., 1923- jt. auth.

174.2 81-65488

Printed in the United States of America

Contents

About the Authors .... **vii**

Introduction .... **xi**

Part I: Fiction Moves Toward Fact **1**

Chapter 1. In The Steps of Hitler **3**
Early attempts at eugenics.

Chapter 2. Experiment in the United States **9**
Eugenics and the United States.

Chapter 3. Some Implications of the Biological Revolution **15**
Genetic tests for fitness in humans?

Chapter 4. Scientists See the Dangers **21**
Genetic engineering—potential for good and evil. Should research be restricted?

Chapter 5. Bizarre Prospects of Cloning **31**
Cloning—some bizarre possibilities. Will the clone be the physical and spiritual embodiment of the donor?

Chapter 6. More About Biological Dangers **41**
Cloning—The Government . . . The Bible.

Part II: Clone Encounters of the First, Second, and Third Kinds **47**

Chapter 7. Clone Encounters of the First Kind: Plants **49**
Totipotent cells and the cloning of carrots.

Chapter 8. Clone Encounters of the Second Kind: Technique in Animals **53**
Cloning in animals—some potential hazards . . . early experiments.

Chapter 9. Clone Encounters of the Second Kind: Frogs **63**
Cloning—how it was done with frogs.
Chapter 10. Clone Encounters of the Third Kind: Humans **71**
Cloning in animals compared to cloning in humans.
Chapter 11. Human Cloning: Investigation of Rorvik's Claims—The Background **77**
The cloning of humans? The beginning of a story.
Chapter 12. ***In His Image:*** **The Story** **85**
The story of the cloning of a human—fact or fantasy?
Chapter 13. A Survey of Biological Progress **97**
Some details about our knowledge of DNA, the stuff of which genes are made.

Part III: Genetic Engineering: A Biological Time Bomb? **105**

Chapter 14. A Survey of Prospects and Dangers **107**
Transferring genetic material from one creature to another—the hazards . . . the potential . . . the limitations.
Chapter 15. Recombinant DNA Research With Bacteria **115**
Transferring foreign genetic material into bacteria—how it is done.
Chapter 16. Recombinant DNA Research In Higher Animals **121**
Gene therapy in humans—hazards . . . potential bene fits . . . limitations.
Chapter 17. Transferring Genetic Material Into Plants **131**
Gene transfer in plants—potential benefits, but can it be done?
Chapter 18. Controlling Our Own Evolution? An Impossible Dream! **135**
The dream dissolves in the face of practical impossibilities.

Part IV: Artificial Insemination and Test-Tube Babies **139**

Chapter 19. Artificial Insemination **141**
Artificial insemination in the human generates legal and social problems.

Chapter 20. "Test-Tube Babies" Are Here **147**
Fertilization outside the human body, but development within—how it is done. Success at last, but potential complications persist.

Part V: Cloning, Creation, and the Golden Age **155**

Chapter 21. A Divine Boundary is Set **157**
Cloning—a perspective from the early chapters of Genesis.

Chapter 22. God Intervenes In His Own Time **163**
The potential for another Tower of Babel

Chapter 23. More Theological and Other Questions ... **171**
The spiritual status of a clone . . . planned modification of man by cloning . . . inbreeding problems . . . reproduction in the marriage bed replaced by cloning in the test-tube?

Chapter 24. The Golden Age is Coming **183**
The Bible reveals the secrets of the abundant life sought vainly by man through his own manipulations.

Part VI: Abortion **189**

Chapter 25. Who Has the Right to Decide? **193**
Birth control . . . sterilization. Abortion: Economic values . . . killing a baby and murder . . . deformed babies . . . Who shall decide?

Chapter 26. Facts About Abortion Are Not Pleasant .. **201**
Methods . . . rape and abortion . . .mental illness and abortion . . . unwanted

pregnancies . . . abortion and medical risks.
Chapter 27. "But It Is So Small!" **209**
The early life of a fetus.
Chapter 28. What the Bible Teaches **213**
Biblical insights on the nature of man . . . taking a life . . . the handicapped . . . miscarriages.

Summary **221**

About the Authors

Duane Gish

After receiving degrees from U.C.L.A. (B.S. in Chemistry) and from the University of California, Berkeley (Ph.D. in Biochemistry), Duane Gish spent eighteen years in biochemical research—at Cornell University Medical College (1953-1956), the Virus Laboratory of the University of California, Berkeley (1956-1960), and with the Upjohn Company, Kalamazoo, Michigan, a pharmaceutical firm (1960-1971). During that time he worked with two Nobel Prize winners and has authored and coauthored about forty technical scientific papers.

For many years Dr. Gish has had a keen interest in the scientific aspects of the creation/evolution controversy. Having become convinced that the theory of evolution is scientifically untenable and that the concept of special creation offers a far superior model for correlating and explaining the scientific evidence related to origins, Dr. Gish left his position in the research division of the Upjohn Company in 1971 to become Associate Director of the Institute for Creation Research, San Diego, California.

Dr. Gish has authored many articles related to creation versus evolution, as well as several books, including the two best-sellers *Evolution? The Fossils Say NO!* and *Dinosaurs: Those Terrible Lizards.* He has lectured and debated on major university campuses throughout the United States, Canada, Australia,

New Zealand, and Britain, as well as lecturing in several other countries, including Holland, Germany, Austria, Denmark, Sweden, and Norway. He is recognized as one of the foremost creation scientists in the world today.

Clifford Wilson

Dr. Clifford Wilson is an Australian scholar who has travelled extensively. For some years he was Director of the Australian Institute of Archaeology. He graduated B.A. and (later) M.A. (Archaeology) from Sydney University, Australia, B. Div., from Melbourne College of Divinity, Australia, and Doctor of Philosophy (Psycholinguistics) from the University of South Carolina. In addition he is a Fellow and Council Member of the Commercial Education Society of Australia, a body under Vice-Regal patronage. One of his books was the best-seller *Crash Go The Chariots.* This was an answer to Erich von Daniken's nonsensical claims as to mysteries of the past being resolved only by astronaut visitors from outer space. Dr. Wilson has written extensively in the areas of archaeology, theology, and psychology. He is academically qualified in each of these areas.

More recently he has been professionally involved in the area of Special Education. For several years he has been on the Faculty of Education at Monash University in Melbourne, Australia. He is a member of the team functioning as Area 7, Special Education, and has "taken his turn" as Area Head. Dr. Wilson has lectured widely to Special Education groups funded by the State Government and has supervised many projects in the area of Special Education.

As a registered psychologist he has been involved in Special Education in his own state, especially in relation to language acquisition in the young child.

He has seen "special" children first hand and

understands many of their problems. Dr. Wilson has learned to appreciate the rights of these young people to engage in the fullest life that is possible. He strongly opposes any scheme of "eugenics" which would lead to the elimination of those members of society who were born with bodies or minds somewhat different from the rest of us.

His strongly held ethical considerations are based on his convictions as a committed, practicing Christian.

Introduction

We are in the midst of a series of revolutions. The silicon chip has revolutionized computer technology. Neil Armstrong's "great step for mankind," when he walked on the moon, reminds us that we have entered the space age. We are no longer earthbound, and our thinking has been revolutionized toward interplanetary travel and space colonies.

Another revolution that has caught the public imagination is in the realm of biology. We hear about test-tube babies, manipulation of our genetic structure, and even about the possibility of cloning our offspring in the near future. Suddenly we are faced with the fact of a surrogate motherhood. It is this suggested potential for biological revolution that interests us in this book.

We shall discuss the revolution itself, attempting to separate fact from fiction and to assess the significance of the facts themselves. We shall consider related medical, ethical, and spiritual questions. We shall ask what relevance the answers have for our society—that in so many ways seems anxious to surpass the progress of thousands of years in less than a decade.

Men have long talked about cloning, not only of humans but of man-animal creatures, chimeras having characteristics of each. One name that comes to mind in the discussion of cloning is Paracelsus, who told of supposed developments of man-like creatures in his laboratory in the early 16th Century A.D. He claimed that men and animals would grow from the same seed, with man making up one category and animals existing in various other shapes and sizes. He argued that women often had sexual intercourse with animals, and that odd monstrosities had resulted from such unions.

Early Myths and Legends

These and other ideas are totally distorted, including concepts involving sodomy, but nevertheless there are many legends associated with people of old times, such as the Greeks (and even earlier peoples), of beings that were part-man, part-animal, or part-bird. Of course, such creatures are impossible, although some scholars believe that genetic manipulation and recombinations might lead to limited combinations of some kinds.

In more recent times there have been other attempts to create "supermen," the best known being that of Adolph Hitler. We shall see that there have been others, even in the United States.

It is an interesting conjecture to ask if Hitler's Nazi government would have used clones if the present stage of development had been reached by his time. The indications are that most certainly such experimentation would have been conducted.

Spiritual Implications

We shall consider spiritual implications, using the Bible as our authority. There we read that God breathed into man the breath of life, and man became a living being, made in the image of God.

Cloning

We shall discuss cloning, the attempt to initiate a developing embryo by transplanting the nucleus from the cell of one individual into the egg from another female individual from which the nucleus (with all the chromosomes) has been removed. In this connection we shall discuss David Rorvik's book, *In His Image*. We are by no means convinced that a child has been cloned—in fact, we do not believe it. However, the potential is there, and Rorvik's title is thought provoking. It is not creation in the true sense, for life is already present.

In the mercy of God the innocent individual thereby formed would presumably have all human potential, including soul and spirit. It is nevertheless a dangerous area into which we have no right to intrude.

People are often tormented by spiritual problems, such as whether or not they have committed the unpardonable sin. Cloning opens the door to other equally dramatic areas of debate. Some will argue that the clone is not truly human. Relevant questions are, "What happens if someone kills him?" "Is it murder?" "Of course it is" . . . "Of course it is not." "Will this new living being have the right to vote?" "Should he gain the various benefits of welfare and all the rest of it?" "Should he be able to enter into a marriage relationship?" Indeed, "Will he be able to procreate by the normal sexual act?"

The questions and answers relating to the moral and spiritual aspects of the biological revolution are likely to have far-reaching implications.

We will show that some cloning experiments with plants and animals have been successful because the selected cells were so young as to be incompletely differentiated. They have been taken from embryo tissue and have not yet completed their development to assume their particular role in the mature body. When we relate this to humans, it certainly would be a macabre point of view to think of cloning from a fetus still in its own early stages of development! Nevertheless, such a possibility has been discussed.

Test-Tube Babies

We shall discuss test-tube babies, and we shall especially consider the work of the English biologist Dr. R. G. Edwards at Cambridge University, with his clinical colleague, Dr. P. C. Steptoe of Oldham General Hospital. They have made it possible for the first "test-tube baby" to be born. Others have followed suit, according to claims being made around the world.

The relevance to cloning is that Edwards and Steptoe have been successful beyond the blastocyst stage (at which point 64 cells have developed) with the implantation of an embryo into the actual mother's own uterus. That was done to circumvent the mother's infertility, resulting from blocked fallopian tubes.

We shall see that there are serious moral problems involved, and there are problems of various other types also associated with these experiments. They touch such matters as individual uniqueness, a concept which would be changed beyond recognition if the process of cloning continues.

Recombinant DNA Research

We shall discuss the potential of recombinant DNA research, that field of genetic engineering which involves transfer of DNA, that is a gene or genes, from one organism to another. This research is designed to develop means of permanently altering the characteristics of plants and animals, in such a way that they assume new functions they had not previously possessed. We will weigh the potential for good of such research versus its potential for disaster.

Gene Therapy

We shall also discuss experiments that it is hoped

some day will make possible the correction of genetic defects responsible for almost 2000 known genetic diseases, including sickle-cell anemia, phenylketonuria, diabetes, Tay-Sachs disease, and others. Does such research really encourage us to believe that practical means will eventually be developed for curing or ameliorating these often fatal diseases?

Abortion

We shall also discuss abortion, considering both the biological and the moral aspects of terminating life at this early stage. Some might consider this topic inappropriate for a book of this kind. However, test-tube babies, cloning, recombinant DNA, and gene therapy are all designed to bring life into the world or to enhance or preserve it, while abortion is designed to destroy it. It thus seems appropriate to discuss one in the light of the others.

Part I

Fiction Moves Toward Fact

Chapter 1

In the Steps Of Hitler

Cloning has come into its own in popular literature. It has captured the imagination of the man in the street. David Rorvik's book, *In His Image: The Cloning of A Man,* made the whole subject a serious talking point in nonscientific circles.

The Decade of The Biological Revolution

Was the child cloned, or was it not? We say "No" and give our reasons. However, that is our answer in 1981, and we are not sure what the answer will be in 1982, or maybe 1983. The 1970's should be known as the Decade of Biological Revolution.

The world has been moving toward that Revolution for many years, involving ideological as well as biological overtones. The idea of a superrace is not new, as we saw so clearly with the efforts of the infamous Adolph Hitler, both before and during World War II.

The concept of the development of the Aryan race, to lead to a superpeople, was taken up by Hitler and his Nazi party in Germany. It had actually been suggested earlier by Frederich Max Muller in 1906. Hitler appropriated the idea, hoping to develop a race that would rule the world. The new breed allegedly would be a superior type whose personal characteristics would naturally predispose them to their appointed roles of leadership.

In fact, the concept goes back even earlier than Hitler and Muller. In *The Republic,* Plato suggested that the best types of animals should be allowed to breed and their progeny developed, whereas inferior types should be discouraged from sexual union and their progeny abandoned.

Hitler's Mass Killings

Hitler practiced this policy, with humans as his "animals." This was especially so with his dastardly mass killings of Jewish men, women, and children, and his eliminating of others, as well, such as Gypsies and mentally retarded people. Thus humans were being depersonalized and regarded the same as animals. The best were selected for breeding experiments, and others were put aside.

Hitler's racial scheme had been outlined, with some measure of caution, by Kurt Hildebrandt back in 1934, in a work whose translated title was *Norm, Degeneration, Disintegration.* He discussed the need for cell activity and elimination in breeding, dealing first with animals, but relating the principles to selective breeding of humans. There were to be negative and positive aspects, but when his policies were applied by Hitler, his theories were more successful with the negative than with the positive. It is one thing *to kill off* millions of supposedly defective humans, but it proved to be quite another *to breed* a generation of supposedly superior Aryans by the selective processes elaborated in his book.

On the *so-called* positive side, Hitler's S.S. men, his elite corps, were to be used as the procreators of the superior race. One of Hitler's key personnel was

Heinrich Himmler. He was responsible for screening the personnel to be involved in the production of that superior race. Both men and women were chosen, and camps were set up in various countries outside Germany, including Norway and Holland. The experiment was a tragic failure. In fact, the offspring had the same number of ailments and shortcomings as other children who were propagated and reared in the usual way.

Hitler's Abject Failure

Hitler's racial experimental camps have been investigated and publicized in the book *Of Pure Blood* by Marc Hillel and Charisa Henry (McGraw-Hill, New York, 1976). The evidence disclosed that the experiment was more than just an abject failure. The rearing of the children deteriorated into a much less than satisfactory arrangement, with all sorts of frictions developing between the supposedly superior women who soon found themselves acting as nursemaids to the children of others, as well as their own.

Many of the children were adopted into other families, and in the post-war years there was a very real effort to forget the whole matter, and allow the children to enjoy a normal upbringing in their adoptive families.

The experiment had continued for approximately ten years until the end of World War II. In addition to the activities with the S.S. troops, over 200,000 young boys and girls of supposedly "right" types of appearance were kidnapped from Poland alone. Many others came from Czechoslovakia, Yugoslavia, and even Russia. Initially they were in camps known as "Lebensborn E.V." The word "Lebensborn" actually means "life source," and the letters "E.V." stand for "registered club."

The first camp was established on December 12,

1935, in the town of Steanhoring in Bavaria, and there were various other camps established throughout Germany and occupied countries. Eventually non-Germans, such as Ukrainians, were accepted into the organization, because so many of the S.S. men had been killed. However, when that happened in 1944 the Lebensborn concept was already in a state of collapse. Hitler's fantastic attempt to produce a super-race had ended in abject failure.

This was true both with the offspring of his *selected* men and women, and with the results shown from the supposedly superior fair, blue-eyed children who had been kidnapped from various parts of Europe. Hillel and Henry tell us that the children who resulted from Hitler's experiment that they could locate 30 years later did not differ in any way as to physique or color of hair or eyes from others who had been born at the same time in ordinary maternity hospitals. Many of the kidnapped children's features changed as time went by, and 30 years later not many of them would have been acceptable according to the original qualifications demanded for that superior race.

So Hitler fought for his perfect race in one way, and it proved to be a total failure. Man's search for the perfect race by genetic engineering will also be a total failure.

Still Three Score Years and Ten

The history of mankind still tells us that, despite all our improvements, we cannot bring our life span to average more than about three score and ten years (70). That figure was laid down as the norm in the days of the great man Moses, more than 3000 years ago, and our life span still stands at about 70 years. Man, created in God's image, is divinely programmed—that applies to the prenatal months, to the developing years of childhood, and to the time of our death. "Miracle cures" might delay our death for a short time, but something will bring us to death's door. Despite man's brilliance and ingenuity, God's decree is still "three score years and ten."

Chapter 2

Experiments in the United States

Hitler was not the first modern leader to advocate a superior race. Nor was experimentation confined to Nazi Germany, though certainly that was the prime area for notorious inhuman practices.

Our modern generation seems to have practically forgotten our own experiments involving eugenics (selective breeding to improve a race), but earlier in this century even prominent people such as Theodore Roosevelt advocated it as an enlightened view of modern Americans. Eugenics was to involve the elimination of those who were biologically unfit—this is called negative eugenics. Positive eugenics was concerned with the increased breeding of those who were found to be physically fit and therefore considered biologically superior types to be further propagated in society.

It is of course relevant to ask who would make the decisions as to the right or wrong types. Many a politician would be unfit and so should be eliminated, but that would not be the practice that would be followed! Rulers would exempt themselves from such barbarity, of course.

Back to the Greeks and Romans

It is an interesting point to notice in passing that the

idea of eugenics goes back to the times of the Greeks. We have already seen that even in *The Republic* Plato urged that in animal reproduction the best of both sexes should be brought together as often as possible, and that the worst should be allowed to come together as seldom as possible. The idea was to improve the quality of the flock.

Julius Caesar likewise offered financial incentives to every Roman mother who would produce a child, and Caesar Augustus later increased the offered reward. However, it is a fact of history that the birthrate of the rich members of Roman society still declined.

In modern times the concept of eugenics was introduced by Sir Francis Galton, a cousin of Charles Darwin. He was a firm believer in his cousin's hypothesis of evolution as outlined in *The Origin of Species.* In his own book, *Hereditary Genius,* he outlined a theory that was supposed to be a practical application of social Darwinism.

It was Galton who originated the term "eugenics," and part of his interest was to find out why there were outstanding persons in his own family tree. He himself started to read at a very early age, and contributed to a wide range of fields by his expertise. He invented the system of fingerprinting for Scotland Yard in England. He himself was exceptionally well-endowed physically and was apparently a person of attractive personality as well.

Eugenics in the United States

For a time, eugenics was strongly advocated in the United States. John Humphrey Noyes was the founder of the Oneida Colony in New York State and was a disciple of Galton. In 1869 he involved 53 women and 38 men in an experiment to breed humans as close as possible to perfection. The women were pledged to become martyrs to science, and this is regarded as the first American experiment in this new concept of eugenics.

It is not our purpose to outline the whole system, but merely to make the point that there have long been those interested in its development, and we shall note its role in America. In the early years of this century there were eugenics societies springing up all over the United States.

One of the early leaders was Margaret Sanger, who later became a leader in a fight with an opposite dimension—she became a great advocate of birth control programs. She became recognized as a feminist, though earlier she had been an outspoken advocate of the idea of the biological superiority of some humans and the inferiority of others. Perhaps it was not such a large leap after all, because she believed that sterilization was the solution for those inferior people (such as the insane and the feebleminded) to prevent breeding. It is not such a great step, after all, from sterilization to the advocacy of birth control . . . and to what some also consider as birth control—abortion. We are not equating birth control and abortion, but it is possible to see connecting links.

Nonsensical Claims for Eugenics

In those early years associated with eugenics, all sorts of nonsensical claims were made. One such claim was that it was possible to decide whether a person was likely to commit a crime or not because of his physiology. A thief was supposed to have thick eyebrows, a crooked nose, and a receding forehead. There were even arguments that sex criminals normally had such things as cracked voices, blond hair, and delicate faces. Murderers were supposed to have cold, glassy eyes, and teeth like dogs.

In the early days of motion picture films, Hollywood tended to use such ideas for their character typing. This all came from the writings of the criminal anthropologist, Cesare Lombrosc, who wrote *The Criminal Man* in 1876. He told us that potential criminals tended to have primitive brains, long arms, scanty beards, flattened noses, and furtive eyes.

Incredible as it may seem, by 1931 the arguments for eugenics had proved so persuasive that in 30 American states sterilization laws had been passed, and many thousands of American citizens had been surgically sterilized.[1]

Changes had to come, of course. Substantial tests were made on military personnel during World War I, and it was astounding to find that the results of these tests indicated that almost half of the recruits were supposed to be feebleminded! The tests also showed that in five of the northern states blacks scored higher than the white recruits had scored in eight of the southern states. As a result, it was suggested (by the white testers!) that the arguments for genetic superiority must at least be tempered by the fact of environmental differences.

As investigations proceeded it became clear that many of the "established cases" of families with supposed very heavy leanings toward crime were not always so well recorded after all. Often the observations were virtually fictional, with judgments made on such issues as the appearance of a father's face that supposedly demonstrated a low mentality.

Eugenics Back in Its Place?

One good result of the developments and increased knowledge as to DNA has been to show that eugenics

---

[1]J. H. Landman, *Human Sterilization,* New York, Mac Millan Co., 1932, p. 259.

must be put back in the corner where it belongs. However, the arguments are not over, and a new form of eugenics has begun to emerge. The argument is that instead of trying to improve intelligence by better environmental conditions, genetic manipulation will be the new way.

A good example is the comment by Professor Charles Frankel, Professor of Philosophy and Public Affairs at Columbia University. He has suggested that genetic manipulation is likely to improve intelligence more cheaply and by more practical means than environmental reforms can do, especially as these latter have proved so disappointing. Frankel says, "There are many mounting signs that the eugenicist's dream of a remade breed, so long in disgrace in consequence of Nazi forays into the field, may be on its way to a comeback."[2]

[2]Charles Frankel, "The Specter of Eugenics," *Commentary,* March, 1974.

Chapter 3

Some Implications Of the Biological Revolution

An article in *Psychology Today* of June, 1974, entitled "The Politics of Genetic Engineering: Who Decides Who's Defective?" by Frederick Ausubal, Jon Beckwith, and Karen Janssen, gives some disturbing information. The National Institutes of Health have funded a $1.7 million study, to be undertaken at the University of Hawaii's Behavioral Biology Laboratory, with the intention of testing 3,200 families who come from Caucasian and Japanese backgrounds. It stated in the grant application that it has long-range significance, because the data will be useful in future decisions relating to "the disturbing but inevitable questions about population control which will have to be made at every level . . . . The purpose of this study is to provide some solid information about the genetic correlates of intelligence so that an informed decision can eventually be made."

Mandatory Genetic Screening

When it is recognized that 46 of the American States have a mandatory genetic screening process to be applied to newborn infants, it is clear that we are indeed well on the way to the national acceptance of genetic engineering and manipulation. There are arguments being put forward for genetic screening before marriage licenses are issued, and there are other far-reaching possibilities. In a 1974 article entitled "The Specter of Eugenics," Charles Frankel quotes Nobel Prize-Winner Sir Francis Crick as suggesting that "no newborn infant should be declared human until it has passed certain tests regarding its genetic endowment . . . and if it fails these tests it forfeits the right to live."

The implications are frightful. One of your authors (Wilson) has had the privilege for several years of being a Senior Lecturer in the Faculty of Education at the Monash University in Melbourne, Australia. In those years he has been a member of the Special Education team, lecturing to students training for their life's work with "special children." Repeated examples demonstrate that many of these children are able to become very worthwhile members of society, but if the sort of argument put forward by Sir Francis Crick had been followed, they would all have been eliminated.

A Mark on the Forehead?

Frankel makes a further quote that is also frightening. He tells of another Nobel Prize-Winner (Linus Pauling) suggesting that a mark should be tattooed on the forehead of every young person, this being a symbol that would show his genotype. According to Pauling, "If this were done, two young people carrying the same seriously defective gene in single dose would recognize the situation at first sight, and would refrain from falling in love with one another." Pauling wants legislation to be passed along those lines.

The argument is nonsensical, of course, for the very fact that they had the same serious defect might itself

Genetic Engineering

MANIPULATING LIFE

Where Does It Stop?

Cloning
Test-Tube Babies
Abortion
Surrogate Mothers

By
Duane T. Gish, Ph.D.
and
Clifford Wilson, Ph.D.

MANIPULATING LIFE: Where Does It Stop?

Copyright © 1981
MASTER BOOKS, A Division of CLP
P. O. Box 15666
San Diego, California 92115

Library of Congress Catalog Card Number 81-65488
ISBN 0-89051-071-7

Cataloging in Publication Data

Gish, Duane Tolbert, 1921-
Manipulating life: where does it stop? [text] / by Duane T. Gish and Clifford A. Wilson.

1. Bioethics. 2. Cloning—Religious aspects. 3. Abortion—Religious aspects. I. Title. II. Wilson, Clifford A., 1923- jt. auth.

174.2 81-65488

Printed in the United States of America

be a point of attraction and of shared experience for those young people. It could actually be a basic cause for their falling in love. In any case, the human ability to fall in love is not limited to the nonpresence of genetic abnormalities, as so many secure marriages between "special" people have demonstrated.

One of the by-products of genetic research is that disorders and various types of abnormalities can be diagnosed before the embryo has reached maturity. On the dangerous side is that a whole series of problems are opened with recombinant DNA, whereby genetic materials are spliced together, even from different organisms, with potentials for creating new life forms.

Cloning would limit heterogeneity,* but there is an argument for recombinant DNA research for biological reasons, such as the investigation of cancer cells. Obviously, there is a great potential for effective diagnosis and consequent possibilities of prevention of disease. It is a Pandora's Box that has been opened, involving both potential blessing and fearful hazards to mankind. It is the likely distortion of scientific progress that is the problem. Cloning, as such, offers little of value for mankind's progress and should be

---

*Each gene is inherited in pairs, one from each parent. When the genes are of the same type, such as the gene for brown eye color, the individual is said to be homozygous for that trait. If an individual has inherited the gene for brown eye color from one parent and the gene for blue eye color from the other parent, the individual is said to be heterozygous for that trait (he will be brown-eyed, since the gene for brown is dominant and the gene for blue is recessive). For some reason not known, the individual that has a high degree of heterozygosity is generally a healthier individual than one who is homozygous for a large number of traits.

put to one side. Recombinant DNA offers potential benefits, but cloning points the way to universal suicide because of the spiritual overtones, as we shall see.

Locating Genes on Chromosomes

It is sometimes claimed that there are many things to be learned in performing experiments in which the cells of two different species are fused. As justification for fusing a human cell and a mouse cell, it is pointed out that when this happens some of the chromosomes drop away. As the fused cell divides, other chromosomes also disappear. It is argued that it is possible to use this technique to determine the function of various genes on the chromosomes that do not duplicate. Thus eventually it would be possible to test the functions of human cells and to find out on which chromosome each controlling gene is located.

As we read the literature, we find that a gene is being described as though it were a machine. Researchers regard it as part of their work to find out how these so-called machines work, how they were originally built, and how they have developed to their present stage. It is commonly believed by the researchers that they will eventually synthesize all the processes, and that as they rearrange genetic patterns, ultimately they will virtually manufacture life itself.

We point out that these extremes are merely wishful thinking. Nevertheless, some believe that the potential is frightening. The wild fantasies of genetic engineers can no longer be pushed to the background as insignificant, so we are told.

A good example is reported by Stanley Cowan in *The Manipulation of Genes.*[1] Dr. Cowan's team, and

---

[1]Published in *Scientific American,* Vol. 233, No. 1, July, 1975, p. 25.

another led by Dr. Herbert Boyer, took two unrelated organisms that were themselves separate, and not able to mate naturally. The two teams isolated the DNA from each, and then hooked them together, giving a new form of cellular life that had not previously existed. Thus the world had hoisted upon it a DNA chimera, and recombinant DNA was born, with all its fearful potential. Where do we go from here?

Chapter 4

Scientists See The Dangers

Those who argue for the right to continue such experiments suggest that ultimately recombinant DNA will itself be a greater force for good than any antibiotics or drugs that humanity has ever known.

Resolving Energy and Other Crises

Not only would basic medical and social problems be resolved, so it is suggested, but there would be all sorts of other advantages as well. Energy needs would be resolved, precious minerals retrieved from the ocean, oil and chemical spills cleaned up, and polution would be dealt with effectively.

On the medical side, it is argued that once we have fuller knowledge of the functioning and sequences of the one hundred thousand (or more) human genes, it will be possible to devise recombinant DNA which will be copied when placed inside a cell, permitting a selective and desired effect on the functions of the cell itself.

According to Howard and Rifkin, ''When this final barrier is breached, we will truly be on the road to the total control of our evolutionary future."[1] Perhaps that is why, as Howard and Rifkin go on to point out, "The Federal Government has become an enthusiastic supporter of the research, handing out at least 180 recombinant DNA research grants to over 80 laboratories around the country in one year alone."[2]

That is not all. Seven major pharmaceutical companies are now engaged in the recombinant DNA race, with a dozen more drug, chemical, and agricultural companies poised to enter this field which is expected to become a multibillion-dollar industry. Large corporations are developing such things as oil-eating microorganisms to clean up oil slicks—hoping that the organisms themselves will not proceed to the oil fields of Texas and elsewhere. There is even an attempt being made to speed up patents so that those who develop these new forms can have exclusive ownership of them.

Research Galloping Out of Control

The stage has been reached where it is recognized that research may be galloping out of control, with an irresistible momentum. Who is to decide whether or not things can proceed? Even if the United States Government were to outlaw these processes, it could not stop developments in England, or in other countries. Dangerous aspects of this work have been suggested. It has been seriously pointed out that the little microorganism that will clean up oil slicks might go on to consume all available oil—not only oil in Texas, but even in our automobiles.

---

[1] *Who Shall Play God?,* Ted Howard and Jeremy Rifkin, Dell Pub. Co., New York, 1977.
[2] *Ibid.,* p. 33.

The fact is, as genes are spliced together from two different species, novel forms of life are being produced (albeit up to now limited to single-celled organisms), but it is not possible to know what their total properties will be beforehand. Once that chimera is placed within a host it will immediately copy itself. It is possible that it would be able to cause, for example, cancer.

It is not hard to imagine the problem that would result if a genetically modified form of the bacteria *Escherichia coli* (the usual host) were carried out of a laboratory on the skin of a laboratory technician, then roaming free, passing cancer on to all sorts of unwilling victims. Some have imagined that cancer epidemics—and many other types of epidemics—would be possible. It is even likely that some of the harmful effects might not be recognized for years, by which time it would be impossible to do anything about it. The Australian press has recently reported an epidemic of flu in dogs, with the serious suggestion that it was caused by a virus escaping from a laboratory. Fiction is all too quickly becoming fact.

What About Government Controls?

There is no way that this can be controlled simply by Government decree. There are advertisements in science magazines which tell of chemicals and various equipments being readily available, and even a high school student could be involved with recombinant DNA right in his own backyard. The dangers are obvious, and the door has already been opened. Biological warfare and all the rest of it are dread possibilities.

The trends are irreversible and, human nature being what it is, it is entirely possible that before many years have passed damaging compounds will have been released on the world with no way to recapture them. Howard and Rifkin quote Dr. Edward Chargaff, Professor Emeritus of Columbia University, as prophesying, "I should say that the spreading of experimental cancer may be confidently expected."[3]

Leading Biologists At First Urged Restraint

It is little wonder that a document was issued on July 26, 1974, from eleven leading biologists urging the biological community to observe a temporary moratorium on the performing of some of the most dangerous of these experiments. The signatories to the document included prestigious names such as James D. Watson, Paul Berg, Cowan, and Boyer.[4]

This led to a temporary delay—but it was only temporary. In February, 1975, 140 biologists from 17 countries met in California, and they decided to get on with their business of continuing research with recombinant DNA. They went back on their own self-imposed ban on basic research, the first that had ever been agreed to in major areas of science.

That conference simply made it clear that the biologists wanted to get on with their research, and they proceeded to resolve that they would do so. Lawyers were present, and they pointed out the legal responsibilities that would ensue if and when researchers created a "biohazard." They were warned it was possible that there could be multimillion-dollar law suits launched against them.[5]

---

[3]Quoted from Lieb Cavaliari, "New Strains of Life or Death?", *New York Times* magazine, August 22, 1976.

[4]This is reported by Paul Berg and others in "Potential Bio-Hazards of Recombinant DNA Molecules," *Science,* Vol. 185, July 26, 1974.

[5]This was reported by Janet H. Weinberg, "Decision at Asilomar," *Science News,* Vol. 107, March 22, 1975, p. 196.

As a result of this warning a two-point safety program was endorsed. By this it was agreed that there would be laboratory precautions, hopefully to provide "physical containment" in relation to potential hazards. They also agreed to utilize "biological containment" by using only an enfeebled *E. coli* to be the host for the recombinant DNA chimeras. It was hoped that this would mean that any escaped chimeras would not be able to live in the natural environment of the world outside the laboratory.

Subsequently the National Institute of Health drafted detailed guidelines, establishing four levels of risk. It is another matter, of course, to decide how these guidelines, which have formally be accepted, will be policed.

Who Can Recall Bacteria?

The fact is that *it is not possible to contain with absolute certainty the very real possibility of a biohazard.* Even in laboratories with special containment facilities, infections are sometimes reported.

Jonathan King's testimony before the Cambridge City Council, on June 23, 1976, is relevant. He is quoted by Howard and Rifkin as pointing out that even in America's Biological Warfare Research Center at Fort Detrick, 435 laboratory-related infections have been documented. That facility has the strictest containment procedures of all the laboratories in the United States. Obviously bacteria will go out from those laboratories, for they make their abode in humans. *Once a recombinant DNA-type bacteria has escaped, there is no way to recall it.* They are not subject to police laws, nor even to the fear of an opposing country.

As we have already stated, even if prohibition were decreed by the Federal Government of the United States, this would not stop experimentation continuing in England, or in other parts of the world. This is not an area where huge amounts of money would be required, as in the development of the early atomic bombs. Even relatively poor countries could continue research in this area. Prohibition would merely postpone the day when a hazardous or even deadly recombinant organism would be created.

Around the world, cancer cells are being involved in the process of fusion with normal cells in attempts to pinpoint the specific chromosomes that are involved in given forms of cancer. Many scientists would argue, with some justification, that if these experiments were prohibited because of the possibility of recombinant DNA being generated, this would involve irresponsibility with regard to their own highly important cancer research.

Who Will Restrain Brilliant Young Scientists?

If one country did order restraints against development, their action could well be neutralized by a neighboring country taking over where they left off. In actual fact, much of the relevant experimental work is being undertaken in England, where the science of cell biology is at a very advanced state. Oxford University is considerably ahead of any American university in the area of experimental embryology. English universities would hardly accept restrictions laid down by U.S. Government authorities.

To say the least, it is most unlikely that brilliant young scientists will keep away from this area of research with its unique challenge. It is virtually certain that experimentation will continue, and it is highly desirable that some of the implications be recognized.

The world should begin to become aware of international consequences, and some sort of international consciousnes should be developed. International agreements might not prevent wars, but at least they put off the evil day for a considerable period of time. At the end of *The Cloning of Man* by Martin Ebon, there is an interesting article entitled "The Future of Genetic Research" by Dr. Joshua Lederberg. It takes the form of a letter to Dr. Ruth Kirschstein, Director of The National Institute of General Medical Sciences at Bethesda, Maryland, dated October 31, 1977. Dr. Lederberg won the Nobel Prize for medicine in 1958 for discoveries concerning genetic recombination, especially the organization of the genetic material of bacteria. He is Professor and Chairman of the Department of Genetics at the School of Medicine of Stanford University in California.

Two Motivating Factors

Dr. Lederberg points out that there are two motivating orientations for future research policy.

1. The alleviation of anguish because of the afflictions clearly labelled as being genetic.
2. The clarifying of the role of genetic factors in a much wider range of diseases, with more complete ethology.

He gives sickle-cell anemia as an outstanding example of Point No. 1. It competes only with Down's Syndrome (Mongolism) in the formidable public health problem that it presents. He says that it is an urgent problem because of the development of practical methods of prenatal diagnosis. He suggests that new methods of study relating to recombinant DNA should be pursued. He points out that we thereby have powerful tools for revealing the specificities of single chromosomes and their role in the inheritance of these complex conditions. He argues that the public should be properly informed about the risks and costs they will suffer from the inhibition of research in this field. He argues against the imposition of "intrusive and bureaucratized control."

"We All Carry Defective Genes"

In passing, we notice also that Dr. Lederberg suggests that the possibility of genetic remodeling of people is mere fanciful talk and is futile even from a purely technical standpoint. He goes on to say that the important penalties of genetic defects are already prevalent, and that it is simply not possible to eradicate them since they continually recur. He makes this comment as to genetic defects being so prevalent: "We all carry *several* such genes in every cell."

That is a highly relevant statement, from a Christian point of view. The Bible teaches that we are representatives of a fallen race, and our bodies carry the marks of that fall. There is no such thing as a perfect human being. There has been only one such perfect being, apart from the original pair, and that One was crucified nearly 2000 years ago. All the rest of us bear marks of the Fall in our bodies. There is no possibility of any perfect human race until Jesus Christ ushers in His spiritual kingdom. As the Apostle Paul puts it, at that time this mortal shall put on immortality and this corruptible body will put on incorruption.

We also have corruptible minds, and it is quite unlikely that the fear of consequences will cause genetic engineering to be put to one side.

Dread Prospects . . .
But Scientists Demand Freedom

Let us summarize. Whether or not we like it, there are hazardous prospects associated with the future activities involving recombinant DNA. Where do we go from here? The bald fact is that if controls become too strict, there is no doubt that research will go underground.

The very nature of science demands freedom. Scientists are not likely to respect the restrictions suggested by concepts of ethics or morality, or by the dictates of national or even international interests. There is often a tendency on the part of scientists to suggest that these are not their problems. As pure scientists interested in research, they are entitled to pursue that research, especially when it can be demonstrated that some benefits are likely to accrue.

Dangers Outweigh Benefits

Undoubtedly some benefits can accrue, but it is also true that the dangers may far outweigh the benefits. The atomic bomb hastened the end of World War II, and there are strong arguments to suggest that nuclear power can be a great benefit to mankind. However, the world is acutely aware that the benefits to a great extent are hidden behind that terror we face, knowingly or unknowingly, with the dawning of each new day.

We not only refer to the possibility of international war, but even to the use of nuclear weapons by terrorist organizations. Hiroshima introduced us to the atomic age, and undoubtedly we have now been introduced to the age of genetic manipulation. The terror that may be introduced, in ways that are totally uncontrollable by the genetic engineers themselves, may be directly opposite to the anticipated benefits. The potential for misuse of this new scientific development is surely real.

Will You Be Brainwashed?

Already there is talk of genetic engineering being used to treat schizophrenia and other illnesses that have psychological or psychiatric bases. Soon it will be possible to alter a person's natural temperament, releasing him from anger, frustration, depression, and all the rest of it—or so it is claimed. It does not take much imagination to realize the possibility, though remote, that before very long the whole community can be effectively brainwashed, so that any "anti-social thinking" will be eliminated. Who will decide what **IS** "anti-social thinking"? It could be determined as being whatever opposes the thinking patterns of a particular ruler or group.

Horror Worse Than Hiroshima?

The preparatory stages are past, and the world is about to see the fruition of a generation of biological discovery. New findings have been taking place—more or less quietly—in laboratories around the world, with the man in the street unaware that he has even now begun to rub shoulders with bacteria-carrying scientists who may possibly be spreading a horror that in years to come will prove to be worse than anything imagined in connection with Hiroshima.

Chapter 5

Bizarre Prospects Of Cloning

Before we proceed further, it is relevant to point out a number of bizarre situations that could develop if cloning succeeds. We stress that human cloning is not yet a possibility, unless we are convinced by David Rorvik's book *In His Image*. (As we proceed, we shall give reasons to indicate that we are far from convinced.) For the moment we put forward some of the grotesque and bizarre possibilities. We stress that only according to popular concepts are these even possibilities. They overlook even elementary aspects—for example the fact that the donor cell would necessarily be immature, not yet specialized, as in an adult human.

For the moment we are allowing our imagination free play, but we have already seen that fiction is fast becoming fact. Let us consider a different possibility.

A woman wants to have a child, but she wants it to be entirely her own. She has the ovum taken from the ovary of her own body, and then has the nucleus of this ovum replaced by the nucleus obtained from a cell from her own body. The ovum contains only 23 chromosomes, half of the full complement of human chromosomes. The other 23 came from the sperm at fertilization. The ovum will not develop into an embryo unless it has the full complement of 46 chromosomes.

The nucleus of the body cell, however, contains the full 46, so she has given to one of her own ova the 46 chromosomes that make up her own being. If this developing egg is placed in her uterus, now her own body will carry an embryo that will develop into a child. (Remember, this is sheer speculation, and we furnish technical data later to show why it is not presently possible.)

That child will have no immediate father. The woman will be the mother of the child, although technically her own body was merely the vessel that saw to it that suitable nutrition, etc., was provided. That was necessary so that the embryo could be brought to full term and be born as a normal baby.

In less bizarre types of cloning, the nucleus introduced into a woman's enucleated egg (an egg from which the nucleus has been removed) might be from a cell taken from her husband, or from one of her children, or from an anonymous donor, known only to an expert in genetics whose job it was to select a donor judged to be genetically fit.

Life Processes Depersonalized

It is reasonable to ask what would become of the concept of the family if widespread human cloning was allowed. What would be the role of parenthood? What about the diversity to which we are accustomed with human beings? Instead of there being nations and races, would there be different brands of clones? What would be the role of women? Would they be expected to bear only one clone at a time, or would that be considered uneconomical? Would they be expected to give birth to five at a time? Some researchers say that there are serious possibilities.

Clone Licenses? And Organ Transplants?

There is even serious talk about licensing cloning. *All this does not necessarily mean that cloning is just around the corner, but it does mean that **in scientific circles the possibility has begun to be taken seriously.***

There are some fantastic sidelights associated with these various aspects of cloning. The woman who had been cloned would have a certain advantage over a male who had one of his cells cloned, for she would have access to her own womb, if necessary, whereas the male would have to hire a willing female to incubate his clone. It would probably have to be just an ordinary human, and not one of those exalted clones!

If sufficient clones were developed, would there be all sorts of interclone possibilities, for example, organ transplants, without ever having to worry about compatibility or rejection? Just as mass-produced motor cars would be transferable, so would organs in the bodies of the clones—or would they? According to some popular writers, this is likely, but in fact each clone would be a separate individual. Would ***you*** unhesitatingly surrender one of your kidneys because it happened to fit the needs of another human who happened to have given one little cell so that you could be mass-produced?

Transferring Pregnancy Problems With Surrogate Mothers

Another possibility would be for a woman who does not want the discomfort of a pregnancy to pay for someone else to bear her child, with the surrogate physical mother being little more than an incubator able to provide a suitable hatching environment. This is discussed more fully in the section on artificial insemination and test-tube babies in Chapters 19 and 20.

Another *supposed* possibility would be for a woman who is past bearing age (and does not any longer produce the life-giving ova) to have a cell from her own body used to donate the nucleus used to artificially "fertilize" another woman's ovum. To make the situation even more bizarre, the implant could be in the body of yet a third woman.

Another bizarre possibility is that the young teenager, ready for "dares," would deliberately produce a replica of himself, especially when the buying price has been reduced, as it inevitably would be. It is entirely possible that the "price" *will* be dramatically reduced, even as we have seen computers come down in price so amazingly in one generation.

Martin Ebon makes the point that living semen can be frozen and shipped across the ocean by airplane and be used to fertilize another living being oceans apart from the original donor. He tells us, "The fertilized egg of a sheep, while incubating in the uterus of a rabbit, has in fact, crossed the ocean and was then retransplanted into another female sheep."

A sheep . . . a rabbit . . . another sheep. Where do we go from here? Will ingenious man be satisfied with simply cloning a human? Will a further step be an attempt to create a part-man, part-animal? It is absurd, grotesque, and unbelievable. And yet—today's science fiction and absurdity is all too often tomorrow's established fact. We are beginning to put together our own "do-it-yourself" kits so that we can produce monsters beside whom the Frankenstein creation would appear to be a benevolent gentleman.

Who Bore My Child?

There are other possibilities. A man could be the

father of a child, using a woman of any race, with a body that is young, or a body that is approaching middle age, with a body that is known to the man, or with a body that is totally unknown.

This all opens the door to a baby market in a way that is totally inhuman and depersonalized. One can imagine legal cases in which the woman whose body felt the stirrings of that babe within her insisting that the child was hers, but the mother who had given just one cell of her body for the planting purpose, demanding that in fact the child was hers.

It is not difficult to imagine law cases between a father and a woman whose body had been used as a mere incubator (or hatchery). She had been paid to endure discomfort over a period of months, and then to go through the pains of childbirth, but then wanted to keep the baby that was and yet was not the product of her body.

Obviously other side-issues are involved. Some of them have been written about at length in science fiction stories for years. One is the possibility of breeding Olympic athletes, to bear all the characteristics of the male or female champion athlete. Then there would be the possibility of breeding geniuses, and the name of Einstein has traditionally been mentioned in this connection, whether or not the concept of him being a genius is true!

The possibility of cloning monsters such as Hitler or Stalin has been put forward to show the horrible possibilities of such processes. Even the idea of reincarnating a dead Pharaoh is sometimes found in the literature—something that is beyond possibility, for cells do indeed die, despite the mythological things that are sometimes written. Great medical men, spiritual leaders, military geniuses, academics from various disciplines—these and many others have been suggested as subjects for cloning. The possibilities are widely discussed, and they have been for years.

Clone My Child!

What of the mother who decided that she wanted a particular child to be reproduced? She has a lovely child, beautiful in appearance. To ensure that the rest of her children are equally beautiful, she clones that particular child. No sexual activity is involved, and there is no father except the father of the first child. Such a happening is entirely within the realm of possibility—if in fact cloning ever comes to reality, as per the popularized versions.

We have said we are talking about bizarre possibilities. Let us take it even further. We have already thought of a mother who finds that one of her children is brilliant, or in other ways is special, and so she arranges to carry the clone of that child in her own womb. Perhaps she has chosen a beautiful child, even though only 13 months old, and decides that is just how she wants the rest of her babies. With cloning—in the future, at a stage of perfection, of course—she now has quintuplets, all at once. They are her own special group, and each of them is without an immediate father. Indeed, in the sense that the "original" of each of those babies would then only be two years of age, they would be without a mother in the true sense. The possibilities are unnatural, impersonal, and detracting from human dignity.

Improving the Human Race?

Many other arguments are put forward for the cloning of a human child. One is that thereby the human species can be improved, as to physical attractions, strength, beauty, and mental ability. It is argued that thereby we could become a race of geniuses, with genetic diseases also minimized. Of course, only the healthy would be allowed to develop. You and I who are ordinary people would not be allowed the privilege of knowing that we would produce a child who would have some of those delightful characteristics of both mother and father.

Another argument is that thereby it would be possible to produce many sets of genetically identical humans, and this would make scientific investigation much easier. One does not see the logic of this argument too easily, however. The "extra" clones would still be individuals, no more ready to be the subject of scientific investigation than any being produced by more usual, if mundane, methods.

Another argument is that by cloning, and other forms of biological engineering, children could be provided for couples who otherwise were unable to produce such a child. It would supposedly also be possible to produce a replica of one who had departed, whether it was somebody famous or somebody who was especially loved, whether a child or a spouse. Such a case might result when a child is accidentally killed. A few still-living cells could be removed from the child's body at or shortly after the time of death and frozen. Later one of these cells could be cloned to produce another child just like the first. Then again, such a method would ensure that the sex of the offspring would be controlled.

Weird and grotesque ideas are also being put forward, such as the possibility of producing embryonic replicas that could be frozen until such times as parts were needed for transplants for the original of that genetic twin separated by one generation.

It's all nonsense, of course. Mass production or not, they ultimately would be separate beings, with no spare eyes or fingers.

Nevertheless, the science fiction rambles on. Clones and more clones. Of course, superclones would be produced for espionage, and so much else that was necessary on the international field.

Psychological Problems for the Clone

Needless to say, the cloned person might well suffer from all sorts of unexpected psychological problems. How would he be sure of who he really was? Would he have a problem much greater than that of the often subordinate identical twin? Would the cloned person simply be the chattel of the original from which he had sprung by the simple donation of one body cell? Would there be a movement to culture cells from famous persons, and indeed would there be a black market of cloned babies? What strange and bizarre opportunities it presents for the teenage (and all too frequently the adult) fantasies of today, with their all-engrossing admiration for rock stars and all the rest!

It is relevant to ask what the cloned being would think of his relationship to God. Would it be a unique relationship, as was possible with every other human being, or would he be expected to be merely in relationship to that human being in whose image he was so clearly made?

Would the "parents" of the clone recognize that the development they aimed at for him was in fact entirely different from that expected of the previous generation? Would those combination of circumstances that made possible the development of a Beethoven, or of other great people, necessarily lead to even the desire for such development on the part of the young clone?

Carry On, Clone?

The argument that the clone would carry on where the donor had left off is almost ludicrous. The donor achieved what he did because of certain environmental conditions, as well as the hereditary factors that gave him certain physical and mental qualities. The clone might have a totally different educational experience, and a totally different physical environment. He might be dramatically different from the original donor in the very areas where, according to science fiction, it would be possible to carry on where the original donor had left off. His motivation might lead him to develop interests entirely different from his genetically identical progenitor. There might be dramatically different tastes, different likes and dislikes, even as at times there are with identical twins.

Some people think seriously about the possibility of cloning because, without recognizing it, man is not only born human but carries an inherent belief in life after death. Archaeological research has produced evidence that man has always believed in two things: the fact of a power beyond himself and of a life beyond the grave.

Firstly, then, he was searching for God or gods, and so he would build an altar to God, or he would make an image of a god. No animal has ever done either of those things. Secondly, the man had within him an inherent something that said he would live on beyond the grave. Thus, he would have his body embalmed, or he would be buried in a tomb that was intended as a home for his life after death. It would have food, wall paintings, furniture, and all the rest of it. There was something in him that said, "My body must not die. I will live on!" Again, no animal ever made such preparations for a life beyond the grave.

When a man thinks of cloning, there is a sense in which he is searching for that life beyond. He himself is human, born with that inherent—but not understood—concept that he is meant to be eternal. That inherent idea is right, for man was indeed meant to be eternal, but sin intruded. However, the way to eternal life is not through cloning, but by accepting the new life available through Jesus Christ.

A man does *not* live on in his clone, except in the sense that any other child would live. Man has only one life. If a clone could be produced, it would be as separate as an identical twin is separate.

Chapter 6

More About Biological Dangers

There are not only grotesque prospects. Examined from many perspectives, there are *also* very real dangers associated with this biological revolution. We are brought to the point where it is no longer enough to say that scientific experimentation and research must be permitted.

Others Have Rights

One may experiment in a way that can harm his own body—if he is so foolish. However, it is an entirely different matter when those experiments are affecting the bodies of others. Recombinant DNA and genetic engineering open the door to just that, and ethical standards should somehow be applied to cause the experiments to cease.

It is not enough to glibly state that scientific development must proceed. We have seen that a strong body of opinion would agree that Hiroshima opened the floodgates to terror in ways that no one wants to repeat. Recombinant DNA and genetic engineering may again open the floodgates—floodgates that are even more dangerous in their potential for evil. This is one time when scientific "advance" should be recognized for what it is. Recombinant DNA and genetic manipulations would be best forgotten.

God put man out of that Paradise of Eden long ago, to prevent him from eating of that mysterious Tree of Life. Despite that, man has defied God through the centuries and has paid the price. In this area of genetic engineering, no doubt man will again defy the principle against eating from the Tree of Life. He can ignore that Divine decree, but the consequences will be on his own head. It will be no good, after the event, to ask, "Why does God allow this?" Man is a responsible being. When he chooses willfully to go against Divine prerogatives, he and he alone must bear the consequences.

However, the sad fact is that when we say "man" we are thinking of only a very small majority of men who have the scientific know-how to do these things. The rest of us are dragged along, and by a law of inexorable consequences, we too must pay the price. The least we can do is to raise our voice in protest.

The Debate Escalates

It is clear that the debate as to the acceptance of the National Institutes of Health Guidelines is not a matter of interest to scientists only. In an article by Clifford Grobstein entitled "The Recombinant DNA Debate," he says, "Yet the policy debate over recombinant DNA research was not laid to rest by the appearance of the NIH Guidelines. Instead, the debate has escalated in recent months, both in intensity and in the range of public involvement."[1]

By this time, as Grobstein points out, the debate has gone beyond the scientist and has become a political issue. He tells us, "Nine scientists at the forum, by

---

[1]*Scientific American,* Vol. 237, No. 1, pp. 22 ff. (1977).

word and deed, reiterated the theme that science has become too consequential either to be left to the self-regulation of scientists or to be allowed to wear a veil of political chastity."

Grobstein summarizes the issues very well. He tells of the possible benefits of such research as more effective and more cheaply-produced pharmaceutical products, a better understanding of the causes of such problems as cancer, better food crops, and wide-ranging approaches to energy problems. On the other hand, there are also the possibilities of worldwide epidemics resulting from newly created pathogens, the triggering of frightening ecological imbalances, the provision of new instruments of terror for rabid groups and militarists, and even the power to dominate and to control the human spirit.

Biological Warfare Renounced

Grobstein points out that the United States is a signatory to an international legal convention that has renounced biological warfare, including research to produce the necessary agents. He also points out that "not all countries have taken this step, and public renunciation without adequate inspection cannot insure that covert activities do not exist. Opponents of recombinant DNA research see its techniques as being ideally suited to serve malevolent purposes, either as agents of organized warfare or of sabotage and terrorism" (p. 31). As Grobstein points out, the techniques do not require large installations or highly sophisticated instrumentation. The potential malevolent applications are obvious.

Grobstein touches another issue that is highly relevant to the subject of this book. He says, "The possibilities of genetic engineering and evolutionary control illustrate the fundamental dilemmas raised by the new capabilities conferred by scientific knowledge'' (p. 32).

No Baby Bonuses . . . But Cloning?

In one sense it is strange that there is any thinking at all about cloning. Around the world there has been talk about the population explosion, with the fear that by 2000 A.D. we will virtually be walking on top of one another. These claims are challenged in many quarters today, especially in Western countries, yet we hear of the possibility of this new process whereby we are supposed to look forward to the development of processes such as cloning which will permit the production of babies by processes which are additional to the natural method.

As we say, *it is strange!* The United States Government does not encourage the bearing of more than two children in the family, and there are no baby bonuses given, as in some other countries, such as Australia. One reason is obvious, of course—Australia needs a greater population, and America does not feel that need. The State of New York has made it legal for any woman who so desires to have an abortion. Other States have quickly followed suit. In most States it is not even necessary for minors to have their parents' consent, and yet here we are faced with the very real possibility of genetic manipulation so that children can be born without the usual love-making or sex involvement.

We have noted grants are available for genetic research. If that was confined to legitimate medical research it would be acceptable. Is funding for potential cloning in that category?

Genetic Engineering in Biblical Revelation?

Genetic engineering can lead to all sorts of dire consequences and diabolical possibilities. Some even believe there are references in the last Biblical book, the Revelation, that involve genetic engineering, cloning, and recombinant DNA. Is that the explanation for beings that have intelligence, but with the bodies of certain animals? Some have argued strongly that this is the case, but we believe this to be impossible. In the same way, it has been put forward that the being who was dead and then made alive again, in that same Book, was in fact a clone. We do not put forward that argument as a point of personal conviction: we merely mention it as the sort of freakish possibility that could be involved with the forms of genetic engineering taking place, at least in embryo, in our generation.

Who Will Own the Babies?

Who will own the new babies if and when cloning comes in? Will the corporations who own the various laboratories have the right to all the life forms they produce in their laboratories? Who will decide the right human types to be licensed, both to produce and also to rear these babies? Will there be attempts made to increase the brain's capacity and in other ways to have a new superrace developed? Will there be other experiments which will alter our internal systems so that we can eat much cheaper foods, such as grass—like cows and sheep out at pasture?

Coming events may cast ominous shadows. We may even now be seeing those shadows around us.

Part II

Clone Encounters Of The First, Second, And Third Kinds

Chapter 7

Clone Encounters Of the First Kind: Plants

David Rorvik has written a book he entitled *In His Image: The Cloning of a Man,* and it has become a best-seller. According to the publicity generated to promote the book, it has been the most extraordinary book of this century. It certainly was an immediate best-seller. It concerned a millionaire who was now old, and wanted to leave an exact replica of himself behind.

In his earlier writings Rorvik had become internationally respected, earning a reputation as a famous science journalist. To say the least, however, his book has been highly controversial. He himself claims that a human clone is a fact, but, as we shall see, there are serious challenges to this claim.

Whether his book is fact or fantasy, it is true that the subject of cloning is very much in the news. It is entirely possible that attempts will be made to clone a human infant in the not too distant future. Indeed, it is more than possible: unless God intervenes, such an attempt is a certainty.

Why do we say, "Clone Encounters of the First, Second, and Third Kind?" The reason is that clones have so far been successful in two lower forms of life that are universally recognized as being in different "compartments" than clones of humans, which would be clones of the third kind. We refer, first, to cloning in plants, and such cloning is no longer new. With some growing things it is possible to break off a small part of the plant or a twig, then plant it, and in due course it will produce a totally separate complete tree or plant. No pollination is involved, and thus, no horticultural sex involvement is necessary. The small green growth was simply cut off and separated from its "parent," then planted. The part that was planted did not even look like the complete plant and flower, being only a small part of it, and yet it contained within itself all that was necessary for that ultimate mature expression of a living thing. We are thus calling clones of this type Clone Encounters of the First Kind. By this process it has been demonstrated that each cell of a plant is so constituted that it can (usually) produce an exact copy of the total plant of which it was originally only a very small part.

Totipotent Cells

As early as the turn of this century, Dr. Gottlieb Haberlandt, an Austrian biologist, was writing about the possibility of plants being "totipotent." By this was meant that each cell could have within itself the potential to duplicate the total plant. An English scholar, Dr. Frederick C. Steward, demonstrated half a century later that Haberlandt was right in his assumptions, and Steward's research and association with others, such as Dr. A. D. Kikorian, have been widely publicized. Dr. Steward was formerly with the University of Rochester, and then later, from 1960 to 1965, he was with Cornell University as Professor of Botany and Director of the Laboratory for Cell Physiology, Growth, and Development. In this capacity he collaborated with others, such as Dr. Kikorian.

By intensive and long sustained teamwork, Steward and his associates worked toward cloning a carrot, commencing not from a twig or a small part cut off from the original, but by taking cells from the root of the carrot. These were put into a mixture of fluids that they thought should have contained all that was needed for their development. They tried salts, sugars, and vitamins, but for a considerable period of time had little success. Eventually they added coconut milk, and instead of the result simply being the red of a carrot, the cells now multiplied rapidly and turned green. Steward reported that progress was unorganized, but the weight of the cells increased 80 fold in 20 days, with 25,000 cells becoming 2.5 million cells.

It is not generally known, but the liquid from the coconut is somewhat parallel to the yolk of an egg. The coconut itself is a nut, and so it is a seed. The white hard interior is equivalent to the white of the egg, while the milk parallels the yolk. It is actually the liquid endosperm of the coconut.

A Carrot is Cloned

Steward continued his research and was eventually able to get beyond simply multiplying carrot cells. He was able to demonstrate that it was possible for this new thing to develop into a normal carrot. Shoots, roots, and flowers were all present, just as though they had been grown normally. Steward had managed to release the potential of the cells so that they were able to demonstrate their own inherent propensities.

This was not a creation of life, but the utilization of the potential that was already in the living materials with which he was working. He was able to cause carrot root cells to grow in a simulated nutrient which gave them the same potential as zygotes. What he had done was to successfully bridge the life cycle of carrots between generations, without the necessity of a "sexual'' act.

There were obviously great practical possibilities from what Steward had done. He had shown that it was possible for carrot cells and other cells to be multiplied like microorganisms and then to be suitably stimulated to grow by the thousands so that they would become facsimiles of the normal plants. He had vindicated Haberlandt's prophecy that such a process could be developed to involve the totipotency of the cells.

That is the first important stage in cloning, and we are calling it an Encounter of the First Kind: the cloning of plants. However, Dr. Steward himself has made it very clear that there is a great difference between the cloning of a carrot and the cloning of any other form of life—especially of human life. Before cloning could take place with any animal it would be necessary for there to be the equivalent of a blood supply from the parent, for that is what the embryo receives in the uterus of the mother. The problems involved with the cloning of an animal or of a human are immensely greater than those associated with the cloning of a carrot. Humans must receive their nutrition from complex sources, and these are distributed by the blood in extremely intricate ways. There is no close parallel to the life that flows through plants, cloned or uncloned.

We shall have more to say about the various details associated with cloning, touching on the technical processes involved, in our next chapter. We shall then go on to elaborate what we mean by ''Clone Encounters of the Second Kind.''

Chapter 8

Clone Encounters Of the Second Kind: Technique In Animals

What *is* cloning in animals? Take one *egg* and remove its nucleus (denucleate it). Then implant into the vacated area the nucleus from a *body cell.* That implanted nucleus has within it the whole genetic blueprint of a mature organism. Theoretically, the egg cell with the implanted nucleus will mature to be an exact replica of the organism from which the nucleus was taken. That new being is a clone. Within that tiny cell there is the genetic information to generate the bodily skeleton, the nerves, the brain, the muscles, the arms, the legs, and all those complex structures that make up a living being.

It is an amazing fact that there are trillions of cells in the human body of about 200 types, and every one of them originally—but only originally—had the potential of all that total being. The cells themselves adapt at a specific period of time for pre-determined uses, and thus obviously the cell that is part of a liver does not become part of an ear. Nevertheless, at first the potential is there, and cloning—even with the cloning of a plant or a frog—demonstrates that there exists a fantastic potential within each of those cells.

Amazing Potential of a Cell

Ultimately this brings us back to the obvious fact that God has created these things, for it is totally beyond the possibility of belief that such potential could be inherent by chance in every single cell of every living being.[1]

There is an absolute and totally powerful Supreme Being behind that life form. Our cells do not have the necessity to use their total capacity, but the amazing fact is that only a very limited use is put on them at all. They serve us as skin cells, or as the cells of our brain, but the potential to do very much more was originally there, inherent in every cell of our bodies.

The more we study, the more we are impressed with the uniqueness of the microscopic entity that we call the cell. Our modern generation has penetrated the cell with fantastic new microscopes. Every individual cell is unique, and popular talk about cloning still retains many concepts that are in the realm of science fiction.

Submicroscopic Markings

It is theorized that every cell in the human body has the potential of acting as the nucleus of an egg. At this stage it is still only theory, and it is not really known what problems will be found as the processes are continued.

It is known that though the "totipotential" is there originally, differentiation (the specialization of the

---

[1]For example see *Speculations and Experiments Related to Theories on the Origin of Life: A Critique,* D. T. Gish, Creation-Life Publishers, San Diego, 1972.

cell's function) takes place at an early stage of development. On those occasions when cloning has been successful, with both plants and animals, the cells used have not been from an adult and have therefore not yet completely differentiated. They have developed from embryonic tissue or larval tissue to complete development, but they were used before they had arrived at that stage of development where they were fully mature.

One important point is, therefore, that it is not really yet known that the cell of a mature man would be able to "act the part" for the total reproduction needed for successful cloning. The most likely possibility is that cloning for a human might be more successful from an embryo than from a fully developed human being. If this is the case, cloning could never serve a useful purpose, since the cell would have to be taken from an embryo, long before it could possibly be known whether or not the individual was superior, average, or inferior.

A highly complex process is involved in removing the cell's nucleus, then inserting another nucleus from another cell. There are microsurgical processes with the fusion operations, involving highly delicate manipulations which utilize difficult techniques. Though the processes could conceivably become relatively simple, we are certainly not at the stage of "backyard cloning," as with "backyard abortion" in so many parts of the world.

Dangers of Damage

The dangers of damage are very great. Each cell contains its own chromosomes. The chromosomes are contained within a nucleus that has its own membrane. Sucking the nucleus out of one cell and inserting it into another cell involves a serious possibility of damage to the nucleus and consequent malformation of the embryo. The difficulties with microsurgery on this scale far exceed those of normal microsurgery, as with stitching in a lens on the eye. Obviously, there is tremendous possibility of damage to the developing organism during the delicate microsurgery required in the possible cloning of a human being.

When we consider the intricate processes involved, it is remarkable to consider the manner in which our own bodies function. One of the amazing facts is that our bodies are constantly replacing cells. Thus the whole process of the division of cells, as well as their multiplication and replenishment, is literally going on throughout all our lives—and without danger to our components. Each of those cells contains the thin thread of DNA that is replicated with further cell division and multiplication.

All this is certainly beyond any possibility of chance development. The functions involved are performed constantly, and with amazing accuracy—and it all happens literally thousands of times a second in the bodies of all human beings.

Rabbits Point the Way

Experiments have also been carried out whereby somatic (body) cells have been fused with unfertilized eggs, especially by Dr. J. D. Bromhall of Oxford who had been a student in earlier times of Dr. J. B. Gurdon. The eggs are denuded, which means that a peeling process takes place to make penetration easier. Then the two cells, one from the body and that of the egg, are put alongside each other. Biochemical prodding takes place, and the cells act as though they have been magnetically brought together, and the body cell's nucleus merges into the egg cell. Their exposure to each other takes about 10 minutes, after which the egg is rolled onto a cell, and it then adheres to the outer membrane before fusion takes place. The success rate was reported as being 18.8%, the nuclei in the eggs being identified after fusion. This was over three times the success rate achieved after surgery.

Bromhall's work was reported in the British journal *Nature,* Dec. 25, 1975. It was especially important because it was dealing with a rabbit, which is a mammal, instead of an amphibian such as a frog.

One interesting and highly relevant point reported by Bromhall was that both cells needed to be at the same stage of comparative development, and must therefore be "synchronized" in development if the experiment was to succeed. Bromhall stated that success appeared to take place only where synchronized donors were transplanted into the eggs. This procedure enabled the eggs to divide as though they were normally fertilized.

Bromhall did not produce a cloned rabbit, but he was able to show that a nucleus from a somatic (body) cell could be transplanted into an unfertilized rabbit's egg and thus take the place of the nucleus produced by union of egg and sperm, at least in the early cleavage state.

It is important to notice that Bromhall stated specifically that the donor and the egg had to be synchronized as to their comparative development. It is reasonable to presume that when this is followed to its logical conclusion, in the supposed cloning of a man, again there would need to be "synchronized" development so far as the transplanted donor cell is concerned.

Each Cell is Unique

We have said that each cell is unique. There is thus no guarantee that the clone physically would be an exact replica of the original. That is illustrated by an experiment with potatoes. In "Science and the Citizen," in *Scientific American,*[2] there is an article, "Test-Tube Potatoes." It tells how James Shepherd and Roger Totsen, of Kansas State University, developed a new means of propagating potatoes. They induced single "naked" cells from the leaves of the mature potato plants to grow in tissue culture, and these would give rise to whole plants. The naked cells (or protoplasts) were released from the leaf tissue by means of enzymes that digested the cell wall material holding the cells together. The researchers stated that when the protoplasts were placed in the appropriate medium, they began to form new cell walls and to divide, yielding a tiny callus of undifferentiated tissue. The article goes on to say that a shoot emerges from the callus, and this eventually develops into a mature plant. Several thousand plants have been generated by this method, this being the first time that the cloning technique has been applied to a major crop plant.

Cloning is a form of asexual reproduction. Therefore the leaf protoplasts of this potato plant should develop into potato plants whose characteristics were exactly identical to those of the parent type. This, however, is not always the case, for the authors say, "Unexpectedly, although many of the clones closely resemble the parental strain, about 25% of them have a physical appearance that is quite different. Some of them have two pistils (the female lower part), some have narrower leaves, or reduced lengths of stem between the leaves, and others look more like climbing ivy than potato plants."

As the article goes on to say, the researchers do not yet understand how it is that these clones have given rise to properties "that are not expressed by the paren-

---

[2]Vol. 238, June, 1978, p. 83.

tal plant, but the process may involve regulatory genes, cytoplasmic factors, or loss of chromosomes in the course of the cloning process'' (p. 84). The article further reports an investigation by Ulrich Matern and Gary A. Strobel of Montana State University. They had collaborated with Shepherd in the isolation of two toxins from the early blight fungus that together elicit the symptoms of the disease called light blight, to which this potato crop is especially liable. ''When they applied the toxins to the leaves of some 500 protoplast-derived clones of a Russet Burbank plant, they noted a wide variation in the resistance of the clones, ranging from the death of the leaves to total resistance.''

The point we are making is, that even with so-called cloning there is no guarantee whatever that the offspring will be exactly the same as the donor. It is a remarkable fact that all cells are themselves individual. Now that we know something of the complexity of the single cell, we should not use the term ''cloning'' in the sense that it is used in popular literature. There is no such thing as cloning in the sense that it is used in that particular popular terminology, for every single cell is uniquely different. In fact, the complexity and intricacy of any one cell is astounding.

Identical Twins and Clones

Those potatoes give witness to the fact that we cannot expect an army of exact replicas, even though they come from one donor. That should be clear from the fact that even identical twins are not totally the same, as has been demonstrated many times by researchers around the world. One of the authors of this book (Gish) has an identical twin, and while he and his twin are indeed very similar, they are not exactly identical. Part of the differences could be due to different influences during development, differences in effect of birth trauma, and different effects of environment, such as differences in effects produced by disease and injuries.

It is sometimes said that identical twins are clones. This is not strictly true, though of course they are the nearest to human clones that we could think of today. There are indeed links between identical twins, for often they suffer from the same illnesses, and sometimes they have even died in similar ways, and very close to each other. There are a remarkable number of identical twins that have married identical twins and have even lived in adjoining apartments. There is often a closeness that is not normally associated with siblings or even fraternal twins. Nevertheless, they are not exact clones.

Identical twins are carried before birth in the same uterus. That is not the situation with the so-called clone where a generation separates the developing embryo from the "original." Also, the body bearing the cloned embryo is totally different from the original of a previous generation. What difference would it make, for example, if one woman bearing the original smoked, and the second bearing the clone did not? Obviously, there are many other possible variables.

"I Take a Dim View of Cloning"

Martin Ebon who wrote *The Cloning of Man,* states categorically, " . . . I take a dim view of cloning a human being. Nor do I believe the claim that such a cloning has actually taken place. The scientific evidence available suggests that experimenters are far from cloning a mammal: the successful cloning of a few frogs seems little more than a fascinating incident on a side-road of genetics. And the achievement of fusing blood cells in rabbits can hardly be regarded as a precedent for cloning a human being."[3]

What do others say? We cannot answer that question until we consider the "cloning of a few frogs." That brings us to "Clone Encounters of the Second Kind."

---

[3] *The Cloning of Man,* Signet, New York, 1978, p. 154.

Chapter 9

Clone Encounters Of the Second Kind: Frogs

We are not setting out to argue about the point of division between plants and animals, or between breeding with plants and breeding with animals. Nor are we trying to separate fish from amphibia, amphibia from reptiles, and reptiles from mammals. However, there is no doubt that there is a major distinction between, for example, a carrot and a frog, or a frog and a human. There are definite distinctions between clone encounters of the first, second, and third types. We saw that carrots can be cloned, and undoubtedly there are commercial prospects as a result of that breakthrough.

"Clone Encounters of the Second Kind" have in their forefront the humble frog. Frogs have been cloned, and to many people this suggests that it stands to reason that humans also can be cloned—that eventually there will be "Clone Encounters of the Third Kind," involving human beings.

Surface Similarities Between Frogs and Humans

To many people there would appear to be great similarities between frogs and humans. However, while it is true that there are similarities, there are also great and complex differences. The cloning of a frog is in a very different category from that of the cloning of a human. Some similarities are obvious. Frogs have four legs, and humans have two legs and two arms, suggesting superficially a basic similarity. Frogs have eyes, and so do we. Frogs breathe on the land and swim in the water, and again there is similarity to most of us humans. Frogs, however, are amphibians, cold-blooded creatures that lay eggs in water or moist environments. The fertilized egg develops into a larva, the tadpole, which has gills and lives in water. The tadpole then undergoes metamorphosis and develops into a mature frog. Humans are warm-blooded mammals, bearing their young alive. Furthermore a frog egg is 1000 times larger than a human egg and thus is much easier to work with. Of course, there are other great differences, as well as similarities.

There have been many attempts to clone frogs, and the majority have been unsuccessful, but there have also been successes. This has provided an important breakthrough. This work has been headed up by Dr. John Bertrand Gurdon. His work has been widely acclaimed, both in England and in the United States. He has been a visiting research fellow a number of times at United States institutions, but his major work has been in the famous English universities, Oxford and Cambridge. His cloning work has been successfully accomplished at the Zoology Department of Oxford University.

He did a great deal of preliminary work during the 1960's and a number of his papers have appeared in the *Journal of Embryology and Experimental Mor-*

phology, from 1970 on. He is widely known for his paper, "The Transplantation of Nuclei from Single Cultured Cells into Enucleate Frogs' Eggs."[1] This was followed by another paper where he described the methods used for the procedures of transplant.

Perhaps it is relevant to point out that the term enucleate in his paper involves the concept of something without a nucleus. The nucleus had been removed, as the yolk might be taken away from an egg.

Gurdon's procedure first involved removing the nucleus from a frog egg. He did this by means of a tiny hollow needle. This left the egg cell devoid of chromosomes. Such an egg, robbed of its genetic material, could not multiply, of course. Gurdon then obtained a cell from some portion of a tadpole of the same species—the intestinal wall, for example. This cell was partially differentiated—it was a cell from the intestinal lining of a tadpole. The cells of a tadpole are not completely and irrevocably differentiated, since the tadpole is the larval stage of a frog, rather than an adult frog.

Gurdon obtained the nucleus from the intestinal wall cell by drawing it up into a pipette with a bore just large enough to admit the nucleus of the cell, but not large enough to admit the complete cell. Thus most of the cell and its contents were stripped away from the nucleus as the nucleus was drawn into the pipette.

Gurdon then inserted the nucleus from the cell of the intestinal wall into the enucleated egg. The egg cell then began to develop, just as would have happened had the original egg been fertilized by a sperm

---

[1]J. B. Gurdon and R. A. Laskey, *Journal of Embryology and Experimental Morphology,* Vol. 24, pp. 227-248, (1970).

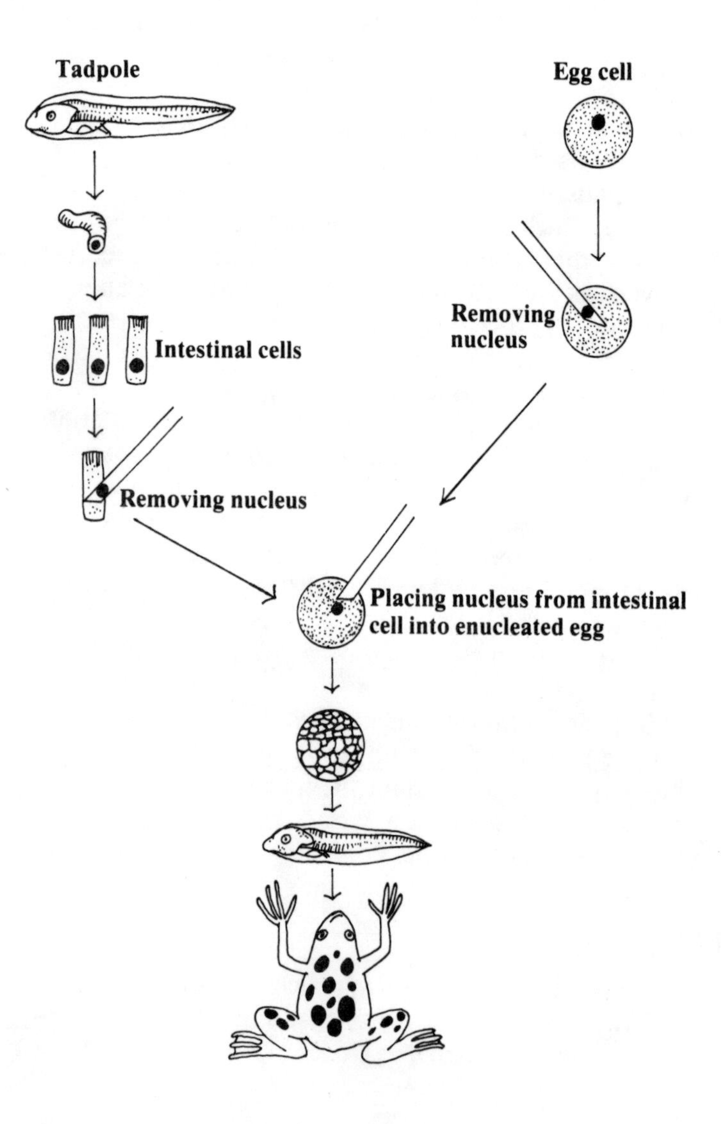

Figure 1. Cloning of a frog. The nucleus is removed from the cell of a tadpole. Development then takes place.

cell. The egg and sperm each carry only half of the number of chromosomes found in somatic (body) cells, such as cells of muscle, brain, liver, kidney, and skin. Thus the human egg and sperm each have 23 chromosomes, while somatic cells (except mature red blood cells which have no nucleus) have 46 chromosomes. After fertilization, which involves union of the sperm with the egg, the fertilized human egg has 46 chromosomes.

The effect of placing the nucleus of the intestinal wall cell in the enucleated frog egg was to give it the same full complement of chromosomes it would have had if the egg had been fertilized in the normal way. In many of these experiments, the egg cell began to develop. Of course, it had no chromosomes from the female frog that had donated the egg. All chromosomes had come from the tadpole. The frog produced in this manner would thus be nearly identical genetically to the tadpole. They would be almost as much alike as identical twins.

Frogs With Twisted Legs.

Gurdon by no means was successful with all his attempts, and he had many difficult experiments before achieving even limited success. It is relevant to point out that even many of the so-called successful experiments were not successful to the fullest extent, for many of the tadpoles that resulted from the experiments had muscle deficiencies, blood problems, and other defects. Some had such eventual complications as twisted legs, but nevertheless, a number of them were "normal," able to feed themselves, and eventually their metamorphosis into adult fertile male or female frogs was complete. It was not an easy process, for Gurdon reported that only 11 clones had been produced out of 707 attempts.

Thus, there was a "Clone Encounter of the Second

Kind.'' This involved a creature about half-way between the single-celled amoeba that can ''procreate'' by dividing itself, and the so-called warm-blooded mammals that include mankind.

Characteristics Are of the Donor, Not the Recipient

It is important to notice that Gurdon was extremely careful to ensure the proper recording of all the technical aspects involved. He showed that it was the nucleus of the donor and not the egg nucleus itself that led to the formation of the transplant tadpoles, proving that the egg nucleus had indeed been removed. Gurdon himself made careful records of the genetic markings of the cells of the cloned frog, and so was able to scientifically establish that the resultant tadpole indeed possessed the characteristics associated with the donor, and not with the recipient.

Gurdon's results are sensational, of course. He made a significant breakthrough, showing that the nucleus from a cell in the body of a creature, such as a frog, could replace the sperm, and could carry its own blueprint of the characteristics of the donor. He has demonstrated that at least some cells can utilize a capacity beyond that which is normally asked of them. Body cells have many varied functions, whether it be in skin, liver, or in the brain, but normally each of them has a specialized use.

The Age of the Cell is Important

We have already said that many results had only relative success. Gurdon found that about three-quarters of the eggs he used would not divide at all, or showed in other ways that the experiment would not be successful in that particular case. There were various defects such as eye damage, twisted legs, and twisted intestines. One conclusion was that there were *many* reasons why some cells would not work, one of which was the age of the cell. Gurdon had to ensure that the cells he used could activate whatever genes were necessary to make possible the great differentiation that would be necessary if there was to be a cloned tadpole, swimming around like other tadpoles produced in the normal way and eventually developing into an adult frog.

Needless to say very specialized instruments were required. The cloning process involved things such as a stereomicroscope, tiny pipettes, and much more. We saw that there was a greal deal of failure, as well as the eventual breakthrough, and as the failures were investigated new techniques were developed.

The view of the cell itself had to be magnified 150 times so that it was possible for the technical aspects to be undertaken visually. A micromanipulator, a type of glass needle, was involved in the process of peeling the cells away from the surface where they had adhered.

Experiments were made with two different types, one being known as the detached cell method and the other as the attached cell method. Gurdon reported that there was no significant difference in the results with each method, though the detached cell method was more efficient in that 100 nuclear transfers would be effected by this method, as against 10 by the attached cell method. Gurdon warns that by this method it was necessary to ensure that the cells were removed from their substrata in such a way that no damage was done.

Chapter 10

Clone Encounters Of the Third Kind: Humans

What about "Clone Encounters of the Third Kind?" Is it really possible for a human to be "cloned"? In the fascinating book *The Cloning of Man* referred to earlier, Marton Ebon has a report by Dr. James Watson. It was presented at the 12th meeting of the Panel on Science and Technology from the Committee on Science and Astronautics of the U.S. House of Representatives, on January 26, 1971. Dr. Watson had already shared the Nobel prize in Medicine in 1962 with two British biophysicists. At the time of his testimony before the House Panel he was Professor of Biology at Harvard University.

Professor Watson briefly outlined the history of cloning, defining the clone as the aggregate of the asexually produced progeny of a single cell. He told how the English zoologist John Gurdon (referred to in previous chapters) had cloned a cell and thereby had settled the question as to whether differentiation in vertebrates was primarily an event that took place as a function of the cytoplasm or of the nucleus. Gurdon had settled the question by showing that differentiation did not involve gene mutations: it was based upon complicated relationships between the nucleus and the cytoplasm, which later effectively commanded certain functions.

Dramatic Differences Between the Three Types of "Clone Encounters"

There are dramatic differences between our three "Clone Encounter" categories, the cloning of (1) a carrot, (2) a frog, or (3) a human or any other so-called higher primate.

Carrots can be produced from a single cell as long as that cell is placed in appropriate nutritional environments. Such environments are readily obtained in ways that would invoke no criticisms whatever, and would not involve decisions of a highly complex and interpersonal nature.

The same is true in the cloning of a tadpole. It is not likely that there would be a "Tadpole Watchers' Society" to protect the rights of the cloned juniors that were swimming around the local pond. It is also a fact that the cloning of a frog is still dramatically different from what would be necessary with the equivalent success in a human. The female frog lays a large number of eggs, and the male frog comes along and deposits his sperm alongside those eggs. Before long fertilization has taken place, and by the normal process of cell division, in due time new tadpoles are on the scene. Both the carrot and the frog have been cloned *in vitro,* outside the body.

Obviously, the differences between these and human cloning are dramatic. There is no likelihood of a human clone taking place in the foreseeable future, apart from those, such as identical twins, which occur normally in the female body.

" . . . Something Like Cloning is Just Science Fiction . . . . "

We quote Dr. Watson: "The chief reason why most

biologists think that possibilities of something like cloning is just science fiction lies in fundamental differences in the embryological development of mammals and amphibians like the frog.''[1]

The frog's egg is relatively large, for it must obviously carry sufficient nourishment—nutrients, to be technical—for the embryological development to take place outside the mother's body. In fact, it must proceed to a point where further nourishment will necessarily be by the embryo feeding itself. This is a major reason why amphibian eggs are large.

With such eggs it is possible to investigate the stage of development relatively easily, because the eggs are outside a living animal's body. They can be put under a microscope and examined by researchers.

This, of course, cannot be done with animals that utilize a placenta inside the mother. Such eggs do not need to be large because of nutritional needs, and in fact they need only sufficient nourishment to reach the 64-cell stage. At that early stage the embryo must be implanted on the wall of the female uterus, and a placenta soon forms. The nutrition for that new life must now come from the female body in which the implanting has taken place.

It is further relevant to notice that Dr. Bromhall is quoted by Martin Ebon as stating categorically that the likelihood of human cloning, either by fusion or any other method, is not possible at this time, and is not likely in the foreseeable future. He also expressed his views that it should not be attempted. Dr. Gurdon had also cautioned that he knew of no true success in experiments with mammals, in that the results of such experiments had not developed normally.[2]

---

[1] *The Cloning of Man,* p. 173.
[2] *Ibid.,* p. 166.

Great Problems With Micromanipulation

As mentioned earlier, there are great problems with micromanipulation. By that process the nucleus is inserted into the egg from which that all-important central portion containing the mother's own chromosomes has been removed.

Nevertheless, techniques have begun to be developed, involving mouse eggs that have been fertilized *in vitro* (outside the body). They are then cultivated under test-tube conditions and developed to the 64-cell stage. The technical name for the development at this 64-cell stage is a "blastocyst." This is the stage past which the embryo must not have developed if it is to be reimplanted in the uterus of the living being for normal development to continue.

We saw that not only mice, but rabbits also have been involved in various experiments. Our interest at this point is only to refer to relative sizes of eggs. One of the remarkable things about the rabbit's egg is that it is 1,000 times smaller than a frog's egg, while the egg of a mouse is about one-third the size of a rabbit's egg. However, the problems of size, in relating the size of such eggs or that of a human egg to that of a frog, are beginning to diminish somewhat because of the tremendous advances with microsurgery.

Other Methods of Fusion

Dr. Watson mentions that other methods have been developed so that two cells can be fused to yield a single cell, but that they would contain the genetic complement of both the donor cells. This is radically different from the basic idea of cloning in the popular sense, which is that the cell would contain the genetic complement of only one donor. That form of cloning would be a sexless procreation, with none of the genetic characteristics of the female bearer of the embryo, even though the embryo was developing within her body. A strange twist would be that such a "clone" could have come from a cell of the woman herself.

The being that is developing in the womb could be either male or female, depending on the donor from whom the one cell was taken. If it was a woman whose cell was used to obtain the nucleus which was placed in the enucleated egg cell, then it would eventually be a baby girl (or a boy if a man's cell was used).

We have given evidence to show that cloning of a human is unlikely in the near future. Nevertheless, a claim of cloning has been made, and we examine that claim in the chapters that follow.

Chapter 11

Human Cloning: Investigation of Rorvik's Claims— The Background

We have referred previously to David Rorvik's best-seller, *In His Image.* It is relevant to state that Rorvik had earlier written in other incredible ways about the possibility of cloning. He had written imaginatively about one hundred clonal offspring to be developed from a man living today, so that 25 years hence the clones could begin the difficult task of colonizing the moon. The selected man was declared the perfect type to serve as a model moon man, and so his clonal offspring would be excellently suited for such a program.

Rorvik as a Science Journalist

Rorvik has been recognized as a reputable scientific writer, and he is widely known for his reports on the activities of men such as Dr. F. C. Steward in the cloning of a carrot, and Dr. J. B. Gurdon, who had successfully cloned a tadpole. Rorvik had anticipated the possibility of shrewd livestock breeders using the new techniques to produce highclass animals. He foresaw that this technique would also be applied to humans, if only to make it possible for an otherwise sterile woman to give birth to a child.

He has written in magazines such as *Esquire*. In April, 1969, that magazine featured his work, with headlines relating to the producing of men and women without the union of men and women; that is, birth without sex. He wrote about imaginative but far-fetched possibilities, such as humans being bred without legs so that they would be more suited for long space journeys, and four-legged men with projecting eyes who would be more suitable for working on the moon.

In his assessment of Rorvik, Martin Ebon tells of his background before he became a well-known science writer. He had planned to write a pornographic science-fiction thriller to be called *The Clone*. It was to be "based on current medical possibilities and perhaps get made into a movie."[1]

It seems that this is exactly what Rorvik has moved toward, for *In His Image* has become a best-seller, even though the publishers acknowledge that they do not know if the "facts" given in the book are truth.

It is relevant to notice that Rorvik had written editorials for the student paper, *The Kaimin,* at the University of Montana. According to Ebon, "Parents threatened to remove their children from the university because of alleged sex-flavored and subversive material published in the newspaper." When Rorvik tried to publish a poem banned from the campus magazine as "flagrantly offensive," the night foreman of the university print shop refused to set the poem in type. Martin Ebon makes it clear that he personally does not accept the story of the supposed cloning as

---

[1] *The Cloning of Man,* pp. 81, 82.

fact—and neither do the authors of this present book.

Rorvik had an outstanding record as a student of journalism. Ebon reports that the Dean of the School of Journalism stated that he had been brilliant, being named as outstanding male graduate of the journalism class at the school in 1966. He was a science writer for *Time* magazine for 1968 and 1969. His ability as a writer is not doubted, but the factuality of his imaginative case of cloning is another matter.

Rorvik is Seriously Criticized . . . And Supported

In the *New York Post* of March 9, 1978, Dr. Beatrice Mintz, who was a member of the Institute for Cancer Research in Philadelphia, referred to Rorvik as being a fraud. Rorvik had included her name on his list of researchers who supposedly had greatly advanced the art of cell fusion. Dr. Mintz said that her work had no connection with cell fusion, and that as far as she knew only one of the scientists whom Rorvik had named was even remotely connected with the concept of cloning. That one exception had attempted to join the egg cell of a mouse with a body cell, but he had not been successful in perpetuating the living creature.

Another figure involved in the controversy was Dr. Landrum B. Shettles who had been at Columbia Presbyterian Hospital. He had successfully cultivated human embryos *in vitro* (outside the body), but none of them had lived longer than a week. Rorvik had collaborated with Shettles in a book called *Choose Your Baby's Sex*. Shettles had moved to Vermont in 1973, after a great controversy that had been written up in the *New York Post*. Shettles had removed an egg from a woman who could not produce a baby, due to blocked fallopian tubes, and had fertilized that egg with sperm from her husband. He had reached the stage where he was ready to place the embryo into the body of the woman. However, a medical officer superior to Shettles at the hospital destroyed the embryo, insisting that further experimentation with animals was needed before such a procedure could be tried on a woman. It is entirely possible that Dr. Shettles' work might have resulted in the first test-tube baby, rather than little Louise Brown having that distinction.

In the book *In His Image: The Cloning of a Man,* Rorvik tells the supposedly true story of a rich man who approached him, prepared to put up one million dollars if his offspring could be cloned successfully.

At the time of the supposed contact, Rorvik tells of contacting Dr. Shettles. Dr. Shettles reported to *Post* magazine that David Rorvik had approached him, with the proposition that he knew of a man who had sufficient funds to pay for cloning if it could be undertaken. Dr. Shettles said that he prepared certain diagrams, but that he knew it would take 18 months to collect information on the equipment he would need. According to the *Post* article, Dr. Shettles suggested that the man must have gotten impatient. He imagined that David Rorvik had gotten somebody else to undertake the operation.

In fairness to Rorvik, it should be said that in the article Dr. Shettles dissociated himself from the argument that the book was a hoax and stated that he trusted Rorvik implicitly. Earlier Shettles himself had claimed that he was well on the way toward developing a so-called test-tube baby. Martin Ebon discusses this and tells of other geneticists who doubted Shettles' claim, nor were they thoroughly convinced by his photographic evidence.

Romantic Ingenuity . . . and Appropriation of a Doctoral Thesis

It is relevant to ask how the doctor who Rorvik

claimed did the work could have succeeded in less time than the 18 months that Dr. Shettles claimed he would need to gather his equipment. It is also relevant to ask how many hundreds of failures there would have been before the supposed clone could have been successfully produced. To say the least, the whole project smacks of romantic ingenuity that is not at a high standard of credibility. For the work to have been successfully undertaken, it would have been necessary to have a team of scientists and others available, and such people are not easy to come by. It is very unlikely that the experiment would have been kept so secret, with no colleague or other person giving the secret information out to a world pressing on the doors, so to speak.

One convincing argument against the genuineness of the whole project is that Dr. J. D. Bromhall of Oxford has taken serious umbrage at some aspects of Rorvik's book. Rorvik himself has extensively quoted Dr. Bromhall, and supposedly his method was used so that the clone could be brought into being. Bromhall had sent Rorvik a nine-page abstract of his doctoral thesis, covering seven years' research.

Rorvik had used these details in such a way as to make a plausible story, but once the story got beyond the point of Dr. Bromhall's survey, the details became blurred and vague. In a letter from Dr. Bromhall to Martin Ebon, this statement appears: "He fails to mention that no cell derived from an adult animal has ever permitted normal development, not even in frogs."[2]

Adult Cells are NOT Satisfactory

We want to emphasize that statement, "No cell

---

[2]*Ibid.*, p. 86.

derived from an adult animal has even permitted normal development, not even in frogs." That should be understood. This flaw in the romantic idea of a man wanting to reproduce himself is not brought out sufficiently in the popularized reports of genetic research. As we have already said, once a cell gets to its state of maturity and is fully developed, it no longer has the capacity to develop in other ways. We have seen that "totipotential" is a factual concept whereby the whole potential of an organism can be in one cell, but that totipotential is no longer available after the ultimate purpose of that cell has been decided. An eye cell will remain an eye cell, and there are reasons to doubt that the nucleus of a cell from a human eye would any longer have the potential to develop into cells that produce legs and arms, fingers, and fingernails, and all the rest. This is a major fallacy with *In His Image* and its romantic idea of a mature adult being successfully cloned.

Dr. Bromhall Files Suit

Dr. Bromhall has filed suit against Lippincott, the publishers of Rorvik's book. He claims damages of seven million dollars because of the unauthorized use of his name, claiming that this was injurious to his own high standing and reputation in experimental embryology. According to Dr. Bromhall, there was no such clone as described by Rorvik, nor does the donor or the experimenter or the surrogate mother that Rorvik describes exist.

Another point that Ebon makes very strongly is that, according to the correspondence, Rorvik was still questioning Bromhall about matters such as cell fusion techniques as late as the summer of 1977, and yet the cloned baby was supposed to have been born in late 1976. Rorvik's factual data turns out to be very weak.

However, we are not only interested in Rorvik's book, *In His Image,* but in the whole Pandora's Box that has been opened by genetic engineering. We shall give some quotes from Rorvik's book and discuss some of its concepts, but we stress that we are not setting out to merely debunk that book. We are considering the frightening implications that will ensue should this field of endeavor succeed. Biological, ethical, and spiritual concepts are likely to be shaken, and the day may come when it will be too late to turn back. We do not know where it will all end, but nevertheless, there is some obligation to raise a voice of protest.

Chapter 12

In His Image: The Story

In *In His Image: The Cloning of a Man,* David Rorvik tells us that a millionaire contacted him. After a series of phone calls and eventual meetings, the story goes, the man told him he would be willing to pay one million dollars if his clone could be produced. The mysterious caller, it is said, had read articles by Rorvik in such magazines as *Esquire* and *Look.* He had also read two books Rorvik had written on the subject of cloning and genetic engineering.

"Max" Was Abandoned as an Infant

Rorvik's story is not all that interesting as a story, but quite often there are some interesting "angles." The millionaire, known as Max, had been orphaned—or abandoned—as an infant and had been passed from one foster home to another. He had not been able to find entirely conclusive records of his own birth, and because of the void in his own life he had developed an exceptional need to have as much conscious will over his own destiny as possible. He did not want his offspring to share that unknown vacuum.

The only way he could die peacefully, according to

Rorvik, was "to first remake himself, in fact to be born again, thus giving himself the willed and wanted and definite origin that he left, or thought he left, under his present circumstances" (p. 20). According to Rorvik, this man "by his own admission [was] the classic over-compensator" (p. 21). He did not want his offspring to have the ruthlessness that had been his own.

One wonders if Rorvik knew full well that there was a serious flaw in his presentation. At page 22 he tells us that Max had a very strong constitution and had never suffered any major illnesses, but more than that, "in terms of his resistance to aging, was regarded as phenomenal by his medical friends." We have already shown that as far as is known, a mature cell could not be utilized satisfactorily, with any hope of eventual success for cloning purposes. This is true whether it be a frog or a mammal.

Later Rorvik gives a seemingly plausible explanation by introducing the idea of taking a new cell from the primary cell. This supposedly by-passed the problem of cell differentiation. We have only Rorvik to quote as our "authority" for such a process. We shall see that his claim fails on other grounds. This is especially so in relation to his still asking for details of procedures (from Dr. Bromhall) ***after*** the baby was supposedly born, but ***before*** the book was published.

Rorvik makes it clear that he has a good understanding of many concepts associated with our genetic structure. For instance, he has Max giving evidence of linguists discovering that there was the potential to utilize a universal grammar included with our genes, and that all of us are born with an innate capacity to comprehend any human language (p. 23). While it is true that there are surface differences between languages, it is also true that children acquire grammar sequentially as appropriate maturational points are reached.

Max Wanted Another Chance "To Be Good"

This amazing inherent capacity with genes was then likened to an inherent capacity for goodness, an area in which Max had been imperfect. Rorvik gathered that Max wanted another chance, through his own clonal offspring, "to be good" (p. 23).

As we read on through Rorvik's book, there are interesting aspects of morality touched on in passing. Thus Max, the millionaire donor, agreed that the details could be made known to the public so long as the anonymity of the party concerned (Max) was maintained, for there had been too much publicity given to quintuplets, and to other multiple-birth children who were made to appear as side-show freaks. Max insisted on anonymity as the one major condition, the breaking of which would be unforgivable.

Rorvik claimed that he did not accept the fee that was offered by Max the millionaire. To do so he would necessarily have had to declare it as income, and he would have been obliged to show who the other party was. Because of Max's insistence on anonymity, Rorvik had no need to fear he would not be compensated if the experiment was a success. The fear of exposure itself would be sufficient to ensure that Max kept his part of the bargain. According to Rorvik, Max would much rather have had Rorvik as an employee, with a hold over him, than as a freelance writer, but with no hold over him.

We are Introduced to "Darwin" And His Methods

The man that Rorvik contacted originally to do the cloning did not go on with the proposal, but he did give another name to Rorvik (so the book tells us), and this opened the door to the alleged cloning of a man. Through the rest of Rorvik's book this man is known simply as "Darwin." Very conveniently, "Darwin" also insisted upon anonymity, and indeed he refused to go on unless there was a firm commitment to this.

We notice as we read further that there are convenient ways of avoiding complex details. He has been accused of giving only such details as he was able to find from scientific experiments by recognized biologists. However, as we read on, we find that he is consistently able to be away from the area of the research for several weeks at a time. This means, of course, that there was no need for him to describe "Darwin's" work or other aspects of what was going on in those weeks. Claims such as he makes—about a baby having been cloned—clearly should be supported by the scientific details.

Except for the minimum of details, for which he has apparently been dependent on input from biologists, Rorvik gets around this lack of information by his convenient absences from the remote site where the cloning took place. Nevertheless, he does give a great deal of technical information which is of interest. He also gives factual material, such as his statement that "the typical woman is born with about half a million egg cells. Yet only about 500 of those would mature in her lifetime" (p. 137). He describes the processes by which eggs were taken from women at this out-of-the-way center. The women were paid well for the minor operations to which they were subjected, so Rorvik says.

We learn (page 139) that the eggs themselves were submitted to all sorts of environmental variables of alkalinity, acidity, osmotic pressures, and ionic and atmospheric factors of different types. There were various experiments to determine the moment in the maturation of these eggs when they would be optimally vulnerable to "fertilization" by body-cell nuclei, accomplished either by fusion or by microsurgery.

They were chemically primed, stretched, denuded of their outer protective layers, pumped up, cooled down, injected, dissected, assembled, disassembled, fused, fertilized, jolted, shocked, irradiated, and sometimes completely dissolved. All these experiments were aimed at getting them to give up their own nuclei and to accept the innards of a "foreign" body cell. Rorvik makes the point that, as far as the body cells were concerned, embryonic cells in which differentiation was not yet complete were ideal. He also tells us that the client in his supposedly factual case had long since left that "larval state" behind him (p. 40).

High Class Science Journalism

As we have already stated, Rorvik had previously been a properly accredited science journalist. Therefore, it is not surprising to find that he gives a great deal of information that is both interesting and relevant in relation to this whole subject of biological engineering. He tells of three different ways by which the biological manipulator (whom he calls "Darwin") could obtain embryos for his experiments.

The first way was to artificially inseminate some of his patients and to perform surgery in order to recover a fertilized egg from their tubes. That however was cumbersome, time-consuming, and even risky. The second process was to produce embryos by utilizing the cloning process, whereby the body-cell nuclei would be implanted into eggs. The third way was to produce the embryos by "test-tube" fertilizing with eggs and sperm. However, in the cloning experiments it was necessary to produce the results not with sperm, but with a body-cell nucleus.

Rorvik describes a whole number of processes such as "capacitation" whereby the heads of sperm are unsheathed by chemical constituents related to the female reproductive tract. He discusses the various developmental stages, such as the "morula," or 32-cell stage, and the "blastula" stage, at which 64 cells had developed. According to Rorvik, "Darwin" had implanted some of his test-tube embryos, and as he goes through the book he discusses the supposed processes of trial and error.

There is at times the romantic and almost the absurd in some parts of Rorvik's book. One example is when he wants to tell us that "Max," though advanced in years, had young cells. This is glibly attributed to his diet. We are then told that Max took various trace minerals, ginseng, and large doses of vitamins A, C, and E. These things were supposedly very helpful in causing Max to have an effective antiaging ingredient in his system.

Another method supposedly used to overcome the problem of the aging of cells was to use the offspring of the original cells, for if the cells were cultured, they seemed less specialized than the parent cells had been, and so were more suitable for nuclei transfer. This is an ingenious way out of the obvious problem that a mature adult could not be cloned because his cells were now specialized.

The last portions of Rorvik's book are relatively uninteresting. He allows the birth to proceed more or less normally, in a small hospital. We are left with the impression that "Max" will eventually marry "Sparrow," the surrogate mother, and that they will live happily ever after.

It is also clear that Max has been furnished with the required evidence that the clone really is his offspring.

Rorvik Philosophizes

According to Rorvik's own statements, his investigations caused him to change many of his previous opinions. He states, "I could no longer view with equanimity the casual destruction of a fetus which increasingly—to me, at least—gave every sign of being a living thing, with a definite will to survive" (p. 28). He watched a fetus kick and struggle for life and had been shocked and revolted at other experiments involving human fetuses. He felt that these things were assaults on us all, and that there might be accumulating problems with the increase of abortion and of experimental feticide.

He elaborated on the fact that research on zygotes (fertilized cells) and *in vitro* ("test-tube") conceptions involved a great deal of destruction of embryonic life. He told of what happened to fertilized eggs that were not used. Cells were examined under microscopes to discover their sex, and those that were not wanted by the parents (who had combined for the possible birth of a child) were "jettisoned." As Rorvik himself puts it, "In plain English, however, 'jettisoning' meant flushing down the drain" (p. 30).

Developed forms of life were jettisoned for various reasons. To Rorvik this was an unconscionable disregard for what might well be genuine human life, somewhat like the old days when unwanted girl babies were left on the tops of mountains to die.

Now he himself, supposedly, was to be involved as a party to the same processes. As he put it, there would most certainly be problems, and "embryos that didn't work out or measure up would be jettisoned" (p. 30). He tells of his own mental problems as he discussed with a biophysicist the difficulty that he faced in making his decision. The advice given was that the act of human cloning being proposed was not necessarily irresponsible or morally reprehensible. Rorvik went along with that, on the face of it, though such a decision seemed to oppose the noble sentiments he had been presenting earlier.

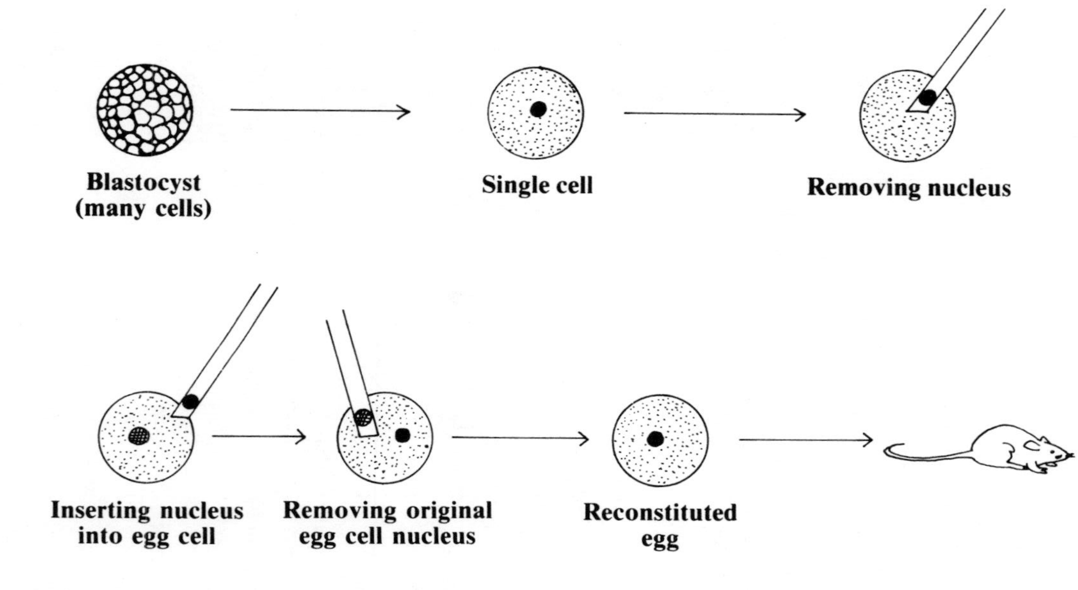

Figure 2. Cloning of a mouse. *The nucleus is removed from a cell obtained from an embryo at the blastocyst stage and this nucleus is inserted into a fertilized mouse egg, after which the original nucleus of the egg cell is removed. The modified egg cell is then placed in a rat uterus where it develops.*

Whose Child Will He Be?

A number of tantalizing questions raise their heads as to the legal status of an anticipated clone: What if the "father" died while the child was still young—would the woman who had borne him have rights to the estate? Would she in fact be the legal guardian of the child, or would it be necessary for the "father" to adopt his own offspring? Would it be possible for the woman who bore him to claim that he was her son, and not the son of him from whom the body cell came? What would be the processes of registration? Would she be declared as the mother of the child?

Another interesting point made by Rorvik is, "It was important that the right amount of cytoplasm be retained. Too much and you interfered with the machinery of the egg: too little and you killed or impaired the functioning of the nucleus" (p. 199).

He also mentioned that they had received information that a research team had successfully used Cytochalasin-B, the substance that could make cells release their nuclei without damaging the cells, in order to enucleate human cells. It was also demonstrated by the same research team that it was possible to put the nucleus into the cytoplasm of another human cell from which the original nucleus had been removed, and then produce new cells that would thrive and grow in culture, at the same rate as undisturbed cells (p. 200).

Rorvik quotes from A. L. Mubbleton-Harris and L. Hayflick, " 'To our knowledge,' these Stanford researchers concluded, 'This is the first time that an isolated karyoplast and cytoplast of a normal human diploid cell have been reconstituted to form a viable, replicating cell.' "[1] (p. 246).

As a matter of fact, the first success in cloning a mammal, very nearly in the sense claimed by Rorvik, was not reported until the manuscript for this book was nearing completion. It was not until mid-1979 that Karl Illmensee and his colleagues at the University of Geneva succeeded in cloning a mouse.[2] The author of the article reporting this development emphasized the skepticism accorded Rorvik's published report because work on simpler animals had not been successfully performed before Rorvik claimed that success had been achieved with humans. Illmensee flatly states that his work represented the first successful nuclear transplant in mammals.

In the three mice that Illmensee has produced, the nuclei were removed from the donors at the blastocyst stage. As will be recalled, the embryo at this stage consists of only 64 cells. Most biologists believe that it is at this stage that cellular differentiation begins.

Illmensee treated the egg cell that was to be the recipient with Cytochalàsin-B. This treatment disrupts the proteins inside the cell that are partially responsible for the rigidity of its gelatinous interior, enabling the cell to withstand penetration by a glass pipette and the insertion of the material from the mouse cell.

Illmensee disrupted the blastocyst to obtain a single cell, the nucleus of which was then drawn into a tiny glass pipette and injected into a fertilized mouse egg.

---

[1]In "Cellular Aging Studied By The Reconstruction Of Replicating Cells From Nuclei and Cytoplasms Isolated From Normal Human Diploid Cells," reported in *Experimental Cell Research,* Vol. 103, (1976), pp. 321-330.

[2]"Cloning Efforts Highlight Research on Mice," J. L. Fox, *Chemical and Engineering News,* Vol. 57, July 30, 1979, pp. 19-21.

The sperm-and-egg nucleus of the mouse egg was then removed and the egg, now containing only the transplanted nucleus, was placed in the uterus of a female mouse. A live and healthy mouse was produced. Illmensee showed that the mouse produced by his procedure possessed the genetic traits of the donor (the embryo mouse) and not those of the mice from which the egg and sperm had been obtained.

You will note that the donor cells did not come from an adult mouse, or even from a new-born mouse, but from an embryo that had developed only to the blastocyst, or 64-cell, stage. Rorvik claimed that in the cloning of Max, adult cells were used.

Illmensee is quoted as stating emphatically that such experiments must not be extended to man. He apparently feels that there is great danger if experiments like these are applied to man. Many malformed and/or mentally crippled babies will result, he fears.

Even though we do not accept all of Rorvik's philosophizing, it is clear that some potentials of such biological engineering are frightening. All this makes it clear that mankind has opened a new door in the realm of biology. Whether or not we like it, an awesome power may be available in the hands of scientists. Unfortunately, one great problem is that new discoveries are frequently abused. What might have been originally thought of as an invention for the good of mankind, over and over again has been turned around and used for purposes of war or other devastation. The potential devastation from biological engineering is terrifying. We agree with Illmensee that cloning experiments should never be performed with humans.

Chapter 13

A Survey of Biological Progress

David Rorvik's *In His Image* is not the only book on cloning. This potential for biological revolution is of such a nature that we will see many more books on this subject in the coming decade. One that has gone into mass circulation is *Who Should Play God?* by Ted Howard and Jeremy Rifkin.[1] They make the claim that the dramatic new scientific breakthrough associated with the discovery of DNA and its workings means that man is reaching a point of dramatic change. They claim that man will be able to irreversibly change the development of species, and to use new forms of "human and posthuman beings" (p. 8).

They tell us that biologists are busy at this sort of work in hundreds of laboratories across the country, and that tens of millions of dollars are being spent in pursuit of the mastery of life (p. 8). At some points they oppose genetic engineering, as with the question of whether biological reengineering should forge ahead with a mass program over the next 25 to 50 years.

---

[1]Dell, New York, 1977.

Producing Novel Forms of Life

Their summary of the new developments is a sobering account, disturbing in the implications for all mankind. They date the commencing time for the new revolution as 1973, with the discovery of recombinant DNA. They rightly make the point that this raises highly significant ethical, political, and social dilemmas, greater than any human society has ever faced before (p. 13).

They point to various apparent desirable processes associated with genetic engineering, such as the possibility of curing some 2,000 monogenic diseases—that is, disorders that have been caused by mutation of a single gene.

They refer to other possible developments, such as the development of new strains of plants, and of superstrains being developed. However, they also tell of other wild schemes, such as the possibility of redesigning human stomachs so that people would be able to utilize cheap hay and grass as food. Yet other researchers are talking about hybridizing humans with lower primates (p. 15). The authors refer to what is taking place as a biological revolution. We consider such schemes not only far-fetched, but biologically impossible.

The Importance of the Work of Watson and Crick

We have seen that actually the revolution started earlier than 1973. In 1953 James Watson and Francis Crick made a formal announcement of the discovery they had made at Cambridge in England. It was published on April 25 in the British science magazine, *Nature*. Their discovery was of great significance in the field of biology. They had uncoiled the mystery of the physical makeup of DNA, which is the fundamen-

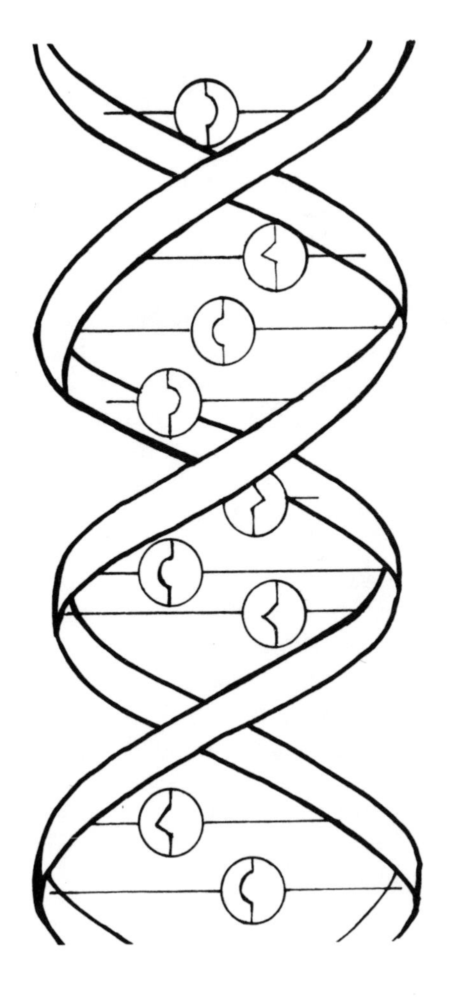

Figure 3. The DNA double helix.

tal genetic molecule.

Watson and Crick were able to demonstrate that DNA is in the shape of a long twisting double helix. It is like a spiral staircase in miniature, having base units, and being composed of four chemical subunits, called nucleotides. These include adenine, thymine, guanine, and cytosine.

The inherent code of the genes themselves determine the kind of proteins that are produced, and the proteins participate in the development of living organisms.

It is now being suggested that if all the DNA in the 30 trillion cells of one human being were unraveled and then stretched out, the strands would reach from the earth to the sun and back some 400 times. The same DNA in all the cells of one human being is so tightly packed that it could fit into a volume that would measure only one cubic inch. This is elaborated by Rick Gore in "The Awesome Worlds Within A Cell."[2]

All the intimate and highly complex programming that is accepted as commonplace for the human is involved in this microscopic data bank, and the biological and chemical functions of life itself depend on these fantastically thin threads of DNA. An amazing amount of information is housed within one single thread of DNA in only one human cell. The color of our eyes, the number of our toes and fingers, the size of our brains, and so much more are all spelled out for us in these tiny threads of DNA.

Some Facts About DNA

1. Biochemists have learned much about how DNA is

---

[2]*National Geographic,* Vol. 120, No. 3, September, 1976, p. 357.

reproduced. The DNA is unzipped, separated into two strands, and then chemical substances from the cell cytoplasm that surround it are attracted and synthesized into DNA. Thus an exact duplicate is made of the original structure. This requires a highly specialized and complex DNA synthesis apparatus found in the cell.

2. We have cracked a small part of the DNA code itself. These developments are elaborated by Lawrence Lessing in his article, "Into the Core of Life Itself," in the magazine *Fortune*, March, 1966, (p. 150 ff.). He tells of experiments by Arthur Kornberg, whose test-tube experiments demonstrated that DNA could indeed be replicated by "unzipping." It was Dr. Marshall W. Nirenberg who in 1961 "performed the biological equivalent of deciphering the Rosetta Stone" by isolating DNA triplets and determining the amino acid they specified. Quickly other biologists have managed to analyze the entire 64-word DNA code.[3] All this was aided by the partial cracking of the DNA code, and the recognition of DNA as a double helix, formed from what are called DNA triplets. The DNA triplets are actually the letters of the genetic alphabet. There are 64 possible combinations that can be derived from the four nucleotides, and each triplet itself codes instructions that are required for the specification of individual amino acids and their sequence in proteins.
3. We have learned much about how DNA transmits its instruction to the cell.
4. We have analyzed a few chromosomes to determine the location of a few genes. It has been known since 1956 that the human cell has 46 chromo-

---

[3] *Who Should Play God?* p. 152.

somes. It is now possible to separate one gene from the rest by laser beam surgery. By these methods scientists are hoping to find out just exactly how each chromosome functions. It is now possible to measure an object 4-billionths of an inch in size, and it has been seriously suggested that one day it will be possible to read off the genetic code along one strand of a chromosome.

5. We have reassembled a cell, with the fusion of the parts of three different amoeba into one whole that actually functions as a unit. In similar fashion, giant cells have been produced which can reproduce themselves to grow between 500 and 1000 times larger than would occur normally. Similarly, at the other end of the scale, minicells have also been engineered.
6. We can now fuse together cells from two different species such as a mouse and a human. This means that the new hybrid cell would carry some of the features of the two different originals. These cells exist only in cell culture and do not develop into an animal.
7. Human genes have been isolated and thus can be analyzed. It is anticipated that this will lead to an increased understanding of the mechanism that actually turns genes on and off.[4]
8. We have "mapped" some genes. This means that we have been able to trace the relative position of genes that are ultimately responsible for various physical traits. It is known that they are located

---

[4]This is elaborated by Jonathan Beckwith in his article "Recombinant DNA: Does the Fault Lie Within Our Genes?" This paper was presented to the National Academy of Sciences Forum on Recombinant DNA, meeting at Washington, D.C., March 7-9, 1977.

in association with specific chromosomes. Over 200 genes have now been mapped.

9. A gene has been synthesized. Dr. Har Gobind Khorana of Massachusetts Institute of Technology, Boston, pioneered the synthesis of genetic material in 1970 when he and his team synthesized a gene. By 1976, it was announced that he had built a 200-unit human gene. He used only the basic four nucleotides, and had added the start and stop mechanisms that would be critical to controlling its function. It was reported that when this was inserted into a cell, the synthetic gene did function.[5]

   It should immediately be pointed out, however, that this functioning was effective only when it was inserted into the cell. This is certainly not "creation" in the true sense of the term. Genetic engineering can go only so far, necessarily requiring the presence of the actual life source itself before being able to function. This is not to say that the results are not important.

   Experiments have been undertaken with newborn mice whereby genetic material has been taken from one mouse and injected into the fetus of another. Genetic material of the first mouse showed up in the newborn mouse. When eventually these mice mated and had offspring, the "foreign" genes were passed on to the next generation.[6]

The title of the book by Howard & Rifkin is, *Who Should Play God?* The facts outlined in this chapter suggest that man has begun to "play God." We shall see in Part IV that he continues to do so at his own peril.

---

[5] "A Working Synthetic Gene," *Medical World News,* September 20, 1976, p. 7.

[6] These 9 points are elaborated by Howard & Rifkin in *Who Should Play God?* pp. 24-26.

Part III

Genetic Engineering: A Biological Time Bomb?

Chapter 14

A Survey of Prospects & Dangers

In this and following chapters we survey some aspects of the research and applications of what has generally been termed "genetic engineering." For the sake of continuity and clarity, some material already mentioned in earlier chapters will be considered in greater detail.

Fears generated by recombinant DNA research have caused outcries among both laymen and scientists, recently resulting in a temporary moratorium on this kind of research. Alarm was expressed over the possibility of converting a harmless bacteria into disease organisms against which there may be no defense. Some dire predictions were made about possible disastrous results that might occur should genetic material be transferred from one animal into another. Efforts were undertaken in some communities to adopt ordinances banning recombinant DNA laboratories.

Alarmist Outcries

Alarmist outcries had been sounded over a decade earlier, inspired by claims voiced by genetic scientists. George W. Beadle, a Nobel Prize winner, had announced in his book, *Genetics and Modern Biology,* published in 1963, that " . . . our knowledge is such that we could, if we chose to do so, direct our own evolutionary future." In an editorial in *Science,* August 11, 1967, entitled, "Will Society Be Prepared?" Marshall Nirenberg, another Nobel Prize winner, stated, "Cells will be programmed with synthetic messages within 25 years."

Immediately after a press conference called by Harvard biologists in 1969 to announce that they had isolated a gene, the *Evening Standard* in London carried the headlines, "Genetic 'Bomb' Fears Grow." On that same day, another London paper, the *Daily Mail,* headlined a story, "The Frightening Facts of Life. Scientists Find Secret of Human Heredity and It Scares Them." Of course these scientists had not discovered the secret of human heredity and they weren't frightened, but the newspaper reporter apparently thought they should be.

Gordon R. Taylor, a science journalist, in his 1968 book entitled *The Biological Time Bomb* characterized new discoveries of biologists "as earth-shaking as the atom bomb." He claimed that results of these discoveries are not going to explode in some distant future, but during the lifetime of many who are living today (some of whom, he believed, may live to be 150 years old!). He mentioned his anticipation of the early possibility of a child being born 100 years after his father's death; human beings conceived and nurtured into life by processes in which sex plays no part; elimination of diseases caused by genetic engineering.

Disasters Are Possible

There would indeed be cause for alarm if a serious possibility exists that a careless or unwitting act of a scientist would unleash upon the public a deadly disease organism against which it had little defense, or if it would become possible some time in the future to control human intelligence, emotions, and personality via genetic engineering. Each could be as disastrous as the other.

Is there a real possibility that before many years scientists will have locked up in their arsenals biological time bombs potentially as explosive as the many hydrogen and atomic bombs that stand poised in the arsenals of the superpowers? What are the potential hazards and benefits of recombinant DNA research? What is genetic engineering?

Yes, there is a possibility that a harmless bacteria might be converted into a disease organism by genetic research. The possibility is slight, however, and containment procedures demanded by present Federal regulations are strict for laboratories operating under potentially hazardous conditions.

Furthermore, such organisms could not harbor genetic material that had not already been around for as long as man has been on the face of the earth. Man would thus possess defense mechanisms against any such organisms as effective as those he now possesses against common disease organisms. On the other hand, the possibility of altering human intelligence and other characteristics in a beneficial way through genetic manipulations is nil . . . and will always remain so. Furthermore, we predict that transfer of genetic information from one mature animal into another mature animal will fail completely, except in those cases where the recipient is a single-celled organism, such as bacteria. As a matter of fact, genetic material from higher animals have already been introduced into bacteria.

Bacteria Have No Defense Mechanisms

Bacteria, being single-celled organisms without specialized organs, do not have the defenses to reject or destroy foreign material. Genetic material from other organisms, both from other species of bacteria and from higher animals, can be introduced into bacteria by special techniques. In some cases this foreign genetic material is incorporated into the genetic material of the recipient bacteria, where it is replicated along with the normal genetic material of the bacteria. In a very few cases it has been reported that the protein coded for by the foreign gene was synthesized by the bacteria.

Procedures similar to the above have succeeded in transferring pieces of genetic material from a higher animal, such as a chick, into cells of another higher animal, such as rat cells. In all such cases, the transfer has not been into the cells of a complete, intact animal, but only into cells grown in cell culture. In these cases, just as with bacterial cells, the cells are single cells. There are no defense organs present to destroy the foreign material. The normal intact animal does, of course, possess organs which provide immunological defenses against foreign substances. When foreign substances penetrate the animal, antibodies against these foreign substances are generated, and specialized cells are mobilized to destroy and engulf the invading substances.

If attempts are made, therefore, to transfer genetic material from one animal to another with the idea of altering some characteristic of the recipient, or of causing it to exhibit some new characteristic altogether, these attempts are doomed to failure. Even in those cases where the transfer of material is from one animal to another of the same species, the material is rejected or destroyed by the recipient unless constant procedures are taken to repress the immunological defenses of the recipient.

Thus when a kidney or heart is transplanted from one human into another, immuno-suppressant drugs must be administered to the recipient throughout the remainder of his lifetime to prevent rejection of the organ. This requirement has an unfortunate consequence, by the way. Repression of the immunological defenses of the recipient also reduces his defenses against other foreign substances, such as viruses and bacteria. As a result, most recipients of organ transplants eventually die of some disease.

Just as the human body recognizes a kidney or heart transplant as foreign or nonself, and thus mobilizes to destroy it, so will the human body or any other animal mobilize its defenses to destroy any foreign genetic material introduced into it, or the products generated as a result of this genetic material. It will thus be impossible to alter humans or animals by transfer of genetic material between creatures of different species.

Recombinant DNA Research Could Produce Scarce Materials

Recombinant DNA research may provide very useful means of using bacterial cells to produce material which is now scarce and difficult to produce. For example, it may be possible to isolate from human cells the gene that codes for insulin, the protein hormone necessary for the regulation of carbohydrate metabolism and the proper level of sugar in the blood. If this gene could then be introduced into bacterial cells and incorporated into the genetic apparatus of these bacterial cells, it is possible that these bacterial cells would begin to produce insulin. It would be anticipated that the insulin could then be isolated from the bacterial cells and used for treating diabetics. Proper scale-up would enable large quantities of the hormone to be produced and would reduce or eliminate the present need of isolating this hormone from the pancreatic glands of animals, such as sheep or cattle.

It may be that some day many of the other protein substances, the availability of which now depends upon their isolation from animal tissues, will be similarly produced. This will depend upon our ability to isolate the fragment of genetic material carrying the desired gene and the ability to find a bacterium that will incorporate this material into its own genetic apparatus and, using it, to actually synthesize the protein coded for by the gene. However, to be practical, such procedures must become more economical than the present method of isolating these substances from animal tissues.

Gene Therapy

Another possible application of recombinant DNA techniques is the treatment of genetic diseases. Juvenile diabetes is the result of the fact that the beta cells of the pancreas of such diabetics do not produce sufficient quantities of insulin, due to some genetic defect. It is suggested that a portion of the appropriate pancreatic cells of such a diabetic be removed and grown in cell culture. Pancreatic cells would be taken from healthy volunteers, or from human cadavers with normal pancreases, grown in cell cultures, and the gene that codes for insulin would be isolated from these healthy cells. This genetic material would then be introduced into the cells of the patient, restoring to them the ability to produce insulin. A mass of these cells would then be implanted in the pancreas of the diabetic patient from which they had been taken and there they would multiply and function to produce the insulin that the patient formerly lacked.

Benefits and Potential Disasters

This procedure, however, may be limited in its benefits and potentially disastrous. The benefit, if any, would accrue solely to the patient. It could not be passed on to his offspring. The genetics of his *pancreatic* cells would have been corrected, but *not* that of his *germ* cells. The patient would still pass along to his offspring the genetic defect that caused his diabetes. There would be only the remotest possibility that all of the sperm-generating tissues of a male could be corrected by such a technique, and it would be manifestly impossible to correct genetic defects in the germ cells of the female.

At birth the ovaries of the human female already possess all the egg cells (in excess of 500,000) that she will ever produce. The ova in the ovaries exist in an immature stage. Normally, each month one ovum in one ovary matures and begins its journey through the reproductive organs. Before its maturation, its genetics have already been established. This took place before birth, having been established at conception and subsequently incorporated into the immature ova during embryological development.

It is impossible, therefore, to use genetic engineering to prevent a female from passing genetic defects on to her offspring. For such treatment to be effective in the male, as mentioned above, all of the sperm-generating tissue in the testes would have to be replaced or else he would be producing some sperm carrying the genetic defect.

It is actually possible that results would be harmful rather than beneficial. We have no assurance, for example, that the required manipulation of these cells would not cause damage or alteration to them. Monsters might be produced instead of healthy individuals. Furthermore, the introduction of genes from an outside source into a cell requires coupling with material that is foreign to the cell. This might produce disastrous results, as will be discussed in more detail later.

Very special techniques must be employed to transfer genetic material into a cell in order for this material to be incorporated into the genetic apparatus of the cell. Simply injecting genes into cells will not cause them to become part of the genetic apparatus of the cell. In order for this to happen, the injected genes must be attached to the chromosomes or must be attached to other genetic material that normally operates within the cell.

Chapter 15

Recombinant DNA Research With Bacteria

Let us briefly consider some of the technical data.

Bacterial Genetics

Bacteria and blue-green algae do not have a nucleus. In these cells, which are called prokaryotes, the genes, or units of heredity, are part of a long strand of DNA (deoxyribonucleic acid). This DNA actually exists in the form of a double strand, one strand being complementary to the other. This double-stranded complementary arrangement is necessary for the reproduction of this genetic material. Only one strand is "read" during transcription of the genetic message for the production of the many activities of the cell. The double-stranded DNA which constitutes the basic genetic material of bacteria and blue-green algae includes many thousands of genes.

This basic DNA material is the material that distinguishes one species of bacteria from another. This is the material that is responsible, for example, for the differences between bacteria, such as *Escherichia coli* and *Bacillus subtilis.* When these cells reproduce, as in cell division, this genetic material is reproduced, and of course, each daughter cell receives a copy.

It has been discovered that bacterial cells also possess other, smaller, strands of DNA. This DNA, also double-stranded, is joined end-to-end in the form of a circle. This type of small DNA unit is called a plasmid. It has been found that resistance to certain antibiotics is carried by these plasmids, the resistance to each antibiotic being due to a separate plasmid. Plasmids may not only be passed from a bacterial cell to its daughter cell during cell division, but these plasmids may be passed to other bacterial cells of the same species which do not possess them.

Resistance to antibiotics is thus an integral part of the genetics of certain bacteria, and may be passed on to offspring or be transferred to other bacterial cells. Resistance to antibiotics thus does not arise by mutation, as had previously been generally supposed. This new evidence destroys the arguments of evolutionists for one of the very few examples of beneficial mutations that they have claimed. It is actually very highly doubtful that there exists a single known example of a truly beneficial mutation. All mutations seem to be in the nature of injuries, which to some extent impair the fertility or viability of the affected organisms.

Transferring DNA Between Bacteria

Two techniques have been devised for transferring genetic material into bacterial cells in such a way that this material becomes integrated into the genetic apparatus of these cells. When cells are infected with viruses to which they are susceptible, the viral DNA is integrated into the regular genetic apparatus (or "chromosome") of the cell and is reproduced along with the normal DNA. Using selective and mild treatments, the virus can be rendered nonpathogenic (no disease is produced), even though the virus can still penetrate the cell and become incorporated into the

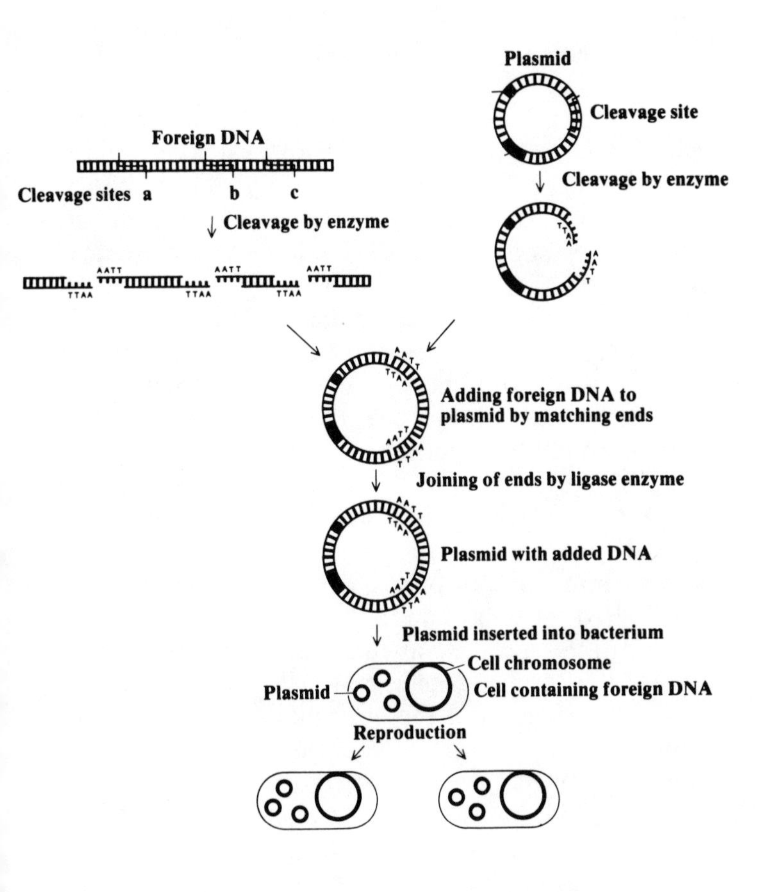

Figure 4. Adding foreign DNA to bacterial cell.

genetic apparatus of the cell. It has been found that when fragments of foreign genetic material are attached to inactivated viral DNA, the viral DNA, along with the attached DNA, can penetrate a cell and become integrated into the genetic apparatus of the cell. As these cells reproduce, they not only reproduce their genetic apparatus along with the viral DNA, but the foreign DNA attached to the viral DNA is reproduced as well.

The other technique utilizes plasmids as carriers. In this procedure, a particular plasmid, for example the plasmid which confers resistance to tetracycline in *E. coli,* is isolated from bacteria bearing this plasmid and which are thus resistant to tetracycline. A special enzyme is used to split this circular double-stranded DNA molecule, resulting in a double-stranded open chain. The same enzyme is used to split the foreign genetic material into pieces, and the piece containing the desired gene is isolated.

Due to the specific way this enzyme functions, each end of the foreign genetic material is complementary to one of the ends of the open plasmid. The foreign gene is mixed with the open plasmid and each end of the foreign gene "finds" its complementary end of the open plasmid, forming a loose circle. Another enzyme is then added, called a ligase, which catalyzes the joining together of the loose ends to form once again a double-stranded circular DNA plasmid which now has been enlarged to contain the foreign gene.

The hybrid plasmid containing the foreign gene is reinserted into *E. coli* cells, none of which were resistant to tetracycline, and thus none of which contained the plasmid which confers tetracycline resistance. Only about one bacterial cell in a million takes up the hybrid plasmid, but these can be easily identified and separated from the other cells because of their newly acquired resistance to tetracycline conferred by the hybrid plasmid. These plasmids do not become part of the basic genetic apparatus of the cell as such (the "chromosome"), but as the cell reproduces itself, the plasmids are reproduced in the cytoplasm and are passed along to the daughter cells.

The foreign gene is reproduced along with the hybrid plasmid, of which it is a part. The bacterial cell has thus been induced to synthesize a gene which has been derived from another organism. In a very few cases, so far, the bacterial cells also manufacture the protein coded for by the gene, including human insulin in one case. The procedure is very complex and tedious, and the final steps in the production of the protein require special chemical processes. Thus the procedure is far from a practical means, at present, for producing proteins on a commercially profitable scale. The procedure potentially has its most valuable application in those cases where the protein product is present in normal tissues in minute quantities, and which is thus enormously expensive. The reader who desires a more detailed description of the above procedures should consult the article in *Scientific American,* Vol. 233, (July, 1975), pp. 24-33, by Stanley N. Cohen, "The Manipulation of Genes."

Chapter 16

Recombinant DNA Research in Higher Animals

The cells of all higher animals and plants possess a nucleus enclosed within a membrane. These types of organisms are called eukaryotes. The nuclei of the cells from these organisms contain the basic genetic material of the cell which is incorporated into the chromosomes during cell division. It has been discovered that certain organelles within the cell also contain DNA. For example, the mitrochondria, the energy-generating centers of the cell, and the organelles of plants which are responsible for photosynthesis, the chloroplasts, each contain DNA material which is involved in the reproduction of these organelles during cell division.

For the introduction of a foreign gene or genes into the cells of higher animals (the eukaryotes), only the route utilizing a virus as the carrier is applicable since the cells of eukaryotes do not contain plasmids. In one experiment, the simian virus, SV-40, was treated with an enzyme that excised a segment of the viral DNA that included the gene that coded for the protein that forms the coat of the virus. The gene for the beta chain of rabbit hemoglobin was obtained and DNA ligase enzyme was used to join this gene to the abbreviated SV-40 viral DNA. This hybrid DNA was then introduced into cells of the African green monkey, grown in culture, and it became integrated into the chromosomes of these monkey cells. It is reported that, among other products, these monkey cells synthesized the beta chain of rabbit hemoglobin. Similar experiments using cells and genes from other animals have been performed.

As we have mentioned in an earlier chapter, if the foreign genetic material were introduced into an intact living animal, the experiment would be doomed to failure. Thus, if the hybrid SV-40 viral DNA to which the gene for the beta chain of rabbit hemoglobin had been attached, were inserted into the intact cells of a living African green monkey, the immunological defenses of the monkey would be mobilized to destroy both the virus and the foreign material attached to it. If some cells did succeed temporarily in incorporating the material and generating proteins coded for by this material, the monkey's body would synthesize antibodies against these proteins, including the beta chain of rabbit hemoglobin.

We thus need have no fear that human beings or other animals will be transformed into strange new creatures using recombinant DNA. God in His wisdom has provided adequate defenses against such attempts. These defenses have been adequate ever since creation to enforce God's decree that each kind was to reproduce only after its own kind and they will continue to suffice to enforce the decree against man's effort to meddle.

Compelling Reasons Against Experimentation With Humans

In addition to the fact that such experiments are doomed to fail anyway, there are compelling reasons why such experiments should never be tried on human beings. The instrument used in these transfer experiments consists of inactive or altered viral DNA, to which the gene is coupled. Although it is very unlikely that such material would result in the induction of the viral disease, since it has been drastically altered (the SV-40 viral DNA used for the cells of the African green monkey, for example, could not code for the viral coat protein), we cannot predict with certainty what results might occur should this material temporarily succeed in implanting itself into the nuclei of some cells of the body.

We could not exclude with absolute certainty, for example, the possibility that such a genetic hybrid would transform these cells into cancer cells.

We could not, of course, exclude other possible harmful results that might be caused by this foreign material. Although it is rather pointless to attempt to imagine just what might happen, we must admit that genetic manipulations of almost any kind in a human body potentially could produce unexpected harmful results. Would we be justified in taking such risks when other therapy is available? Would you wish to risk the life or the health of your diabetic child in a genetic engineering experiment when he already is being safely treated by insulin injections? Experiments of this kind could only be justified in those cases, such as sickle-cell anemia or Tay-Sachs disease, where no therapy is available, and the child is otherwise certain to die anyhow.

In order to clear up any possible confusion, let us review the different types of genetic engineering we have considered so far. The genetic engineering we have just discussed above, as in the case of sickle-cell anemia or Tay-Sachs disease, is designed to correct a genetic defect and to restore the individual to the normal condition.

Sickle-Cell Anemia

In the case of sickle-cell anemia, a mutation in the gene that codes for hemoglobin, the protein of the red blood cell, results in a change in the chemical structure of the hemoglobin. Hemoglobin is a complex that includes four protein molecules. Two of these consist of a protein called the alpha chain and two consist of a protein called the beta chain. The alpha chain contains 146 amino acids (there are 20 different kinds of amino acids in proteins), and the beta chain consists of 141 amino acids. As a result of the mutation (a mutation is an alteration in a gene), the amino acid at position six in the beta chain, glutamic acid, is replaced by valine. Changing only one amino acid out of 287 (146 in the alpha chain and 141 in the beta chain) does not seem like much, chemically speaking, but the biological results are disastrous, usually resulting in death within a few years of birth.

The change of the single amino acid in the beta chain of hemoglobin drastically diminishes the ability of the red blood cell to bind and transport oxygen. In addition, the cell assumes the shape of a sickle, thus the term, sickle-cell anemia. In this case, genetic therapy would consist of an attempt to insert the normal gene for hemoglobin production into the defective cells of the patient.

Gene Therapy for Sickle Cell Anemia

Some of the tissue that generates red blood cells would be removed from the patient and grown in cell culture. The genes that code for hemoglobin production would be obtained from a normal healthy individual and inserted into the defective cells of the patient.

It would first be necessary to attach this genetic material to an inactivated virus, as described earlier, in order for it to become incorporated into the genetic apparatus of the cells. These cells would not contain the gene for normal hemoglobin. After a crop of these cells had been generated in cell culture, they would be placed in the patient from whom the parent cells had been obtained. If they were successful in being incorporated into the red blood cell generating tissue, the patient might now produce sufficient normal hemoglobin to survive.

Since, however, it was necessary to attach the normal gene for hemoglobin to an inactivated virus in order for this gene to be incorporated into one of the chromosomes of the cell, the cells so produced not only contain the normal hemoglobin gene, but they also contain the DNA from the inactivated virus. This viral DNA will be reproduced along with all the normal genes of the cell and will thus be found in every cell so produced. It is the presence of this viral DNA that might produce disastrous results.

As long as these cells are merely being grown in cell culture, the presence of the viral DNA might not produce any harmful effects, since it has been treated to render the virus noninfective. Furthermore, these individual cells possess no defense mechanisms that would be available to recognize and destroy foreign material. When these cells are implanted in the patient, however, the patient does have organs and tissues designed to recognize and destroy foreign material. The introduction of the cells containing the viral DNA into the patient thus might generate antibodies against the viral DNA, and other defenses, such as white blood cells, might also be marshalled to destroy the cells containing the viral DNA. Thus, it might not be long before all of these cells were destroyed and the patient was back where he started.

If, however, as we have mentioned earlier, the patient's immunological defenses are so weak that the inserted cells with their foreign viral-derived DNA manage to survive and multiply in the patient, this viral DNA might transform the cells into cancer cells, or might have some other harmful effect. Thus, rather than the anticipated beneficial effect, the result for the patient might involve a much more serious disease and even death. As we have said, such experiments could only be justified in cases such as sickle-cell anemia, where the patient is almost certain to die without some such heroic effort.

Will Gene Therapy Be Self-Defeating?

Even in such cases as sickle-cell anemia, however, medical science may face a dilemma. If genetic engineering can be used to cure or ameliorate the anemia of sickle-cell victims, the patient will survive and reach reproductive age. If the patient then marries and produces offspring, all of his offspring will either be carriers of the sickle-cell trait or will have the disease.

The gene for sickle-cell anemia is recessive and the gene for normal hemoglobin is dominant. Thus, if an individual inherits the gene for sickle-cell anemia from one parent and the gene for normal hemoglobin from the other parent (all genes for each trait occur in pairs, one gene coming from each parent), he will produce sufficient normal hemoglobin to be healthy. The gene for normal hemoglobin is dominant, essentially preventing any ill effects of the sickle-cell gene. If an individual inherits the gene for sickle-cell anemia from each parent, he cannot produce any normal hemoglobin, and usually dies within a few years of birth.

If, by genetic engineering, the sickle-cell anemia of the patient could be cured or ameliorated, the patient would, as mentioned above, eventually marry and would likely have children. If his marriage partner possesses only the gene for normal hemoglobin, and he possesses only the gene for sickle-cell anemia, each of the offspring will possess a gene for normal hemoglobin and a gene for sickle-cell hemoglobin.

They will thus all be carriers of the disease, but will themselves be healthy, since the gene for normal hemoglobin is dominant. If the patient's marriage partner, however, is a carrier of the sickle-cell trait—that is, if she has a gene for normal hemoglobin and gene for sickle-cell hemoglobin—half of their children will be carriers and half will have sickle-cell anemia, as shown by the following diagram:

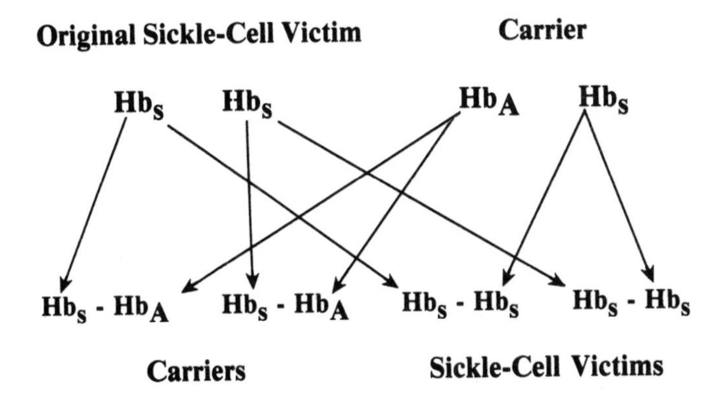

(Hb_S: sickle-cell gene; Hb_A: gene for normal hemoglobin)

Genetic engineering, if successful, will thus save the life of the sickle-cell patient, but will impair the genetic health of the population. If the patient marries a partner with normal genes, all offspring will be carriers of this genetic disease. If he marries a partner who is a carrier, half of the children will be carriers and half will have the disease. This will have the effect of increasing the incidence of the disease, since both the number of carriers and the number of immediate victims would be increased. Without genetic engineering, all or almost all sickle-cell anemia *victims* die before they reach the age of reproduction, and thus do not pass this harmful gene on to the next generation. However, you will recall that the sickle-cell gene is not dominant, therefore, the *carriers* continue to mature and reproduce, passing the harmful gene on to their children. Thus, the disease is never completely eliminated.

The Dilemma

The dilemma is this: shall efforts be taken to save the lives of sickle-cell victims, even though this will inevitably result in the production of more sickle-cell victims? Of course, prospective marriage partners for sickle-cell patients on whom genetic engineering has been performed could be screened to eliminate all carriers of the sickle-cell gene. This would result in a rather embarrassing situation, however, for all proposals for marriage would have to be conditional. The person to whom the proposal is given, if the proposal is accepted, would have to be asked to submit to a genetic test for the sickle-cell gene, and if she failed the test the proposal would have to be rescinded. If the woman to whom a proposal is given was one on whom the genetic engineering had been performed, she would accept the proposal only on the condition that the would-be bridegroom submit to and successfully pass a genetic test.

As another alternative, it is strongly advised that two carriers of sickle-cell seek genetic and spiritual counseling prior to marriage to consider the possibilities of adopting, as opposed to personal procreation. There are many loveable children already in this world —not by their own choosing— who need someone to *choose them* to love and care for.

In any case, the number of carriers of the sickle-cell gene would increase, and the incidence of sickle-cell anemia would thus increase unless the entire population (of blacks, since only blacks are victims of this particular genetic mutation) were genetically tested, and marriage between carriers were prevented. This same factor would be at work in all cases of genetic diseases. Genetic engineering, if this ever does succeed (and that is far from certain), will inevitably impair the genetic well-being of the population even though it would be of immediate benefit to the victims of these genetic diseases. More and more victims of genetic diseases would be generated, requiring more and more genetic engineering, generating yet more and more victims. Eventually, the genetic load would overwhelm medical and economic resources. Genetic engineering, in the long run, might be self-defeating.

Prospects Are Dim

The prospects for genetic engineering as a means of treating genetic diseases thus appear to be very dim indeed. The technique required is complex, and success in obtaining and transferring the genetic material is difficult to achieve. Target cells must first be obtained from the patient, grown in cell culture, inserted with the genetic material, and then reimplanted into the patient. Since viral DNA would be present in the treated cells, they may be attacked and destroyed by the patient's immunological defenses. The presence of the viral DNA might transform the cells into cancer cells, or have some other unforeseen disastrous results. Benefits, if any, will favor only the patient, since his offspring will still inherit the defective gene. The latter fact may eventually result in the population accruing an intolerable genetic load, finally overwhelming medical and economic resources.

The second type of genetic engineering we have discussed is the transfer of genetic material between species. As mentioned earlier, this has already been accomplished where the target cells were single-celled organisms, such as bacteria, or where the target cells were cells from higher animals grown in cell culture. If such recombinant DNA experiments succeed in inducing bacteria to produce important products such as insulin, growth hormone, interferon, or other difficult-to-obtain materials, this technique may prove to be of considerable benefit.

On the other hand, as we have emphasized in our earlier discussion, all attempts to insert into an intact animal genetic material that has been obtained from another species are doomed to failure. Such insertion will immediately cause the immunological defenses of the animal to be mobilized to destroy the cells containing this foreign material, in the same way that these defenses are mobilized to destroy invading bacteria, viruses, and transplanted kidneys. The idea that scientists will be able to create chimera by transferring genetic material between different species seems to be contradicted by all presently available evidence.

Chapter 17

Transferring Genetic Material Into Plants

Another possible application of genetic engineering is that which might be applied to plants. If it were possible to transfer foreign genes into plants, several useful applications immediately come to mind. Plants do not have the ability to fix nitrogen (nitrogen fixation involves absorbing nitrogen from the air and converting it into nitrogen-containing compounds such as ammonia).

Certain plants, the legumes, establish a symbiotic relationship with bacteria, such as *Rhizobium,* which do have the ability to fix nitrogen. When *Rhizobium* invade these plants, the bacteria supply the plants with nitrogen-containing compounds in exchange for nutrients from the plant. If the genetic material that is responsible for nitrogen fixation in these bacteria could be placed directly in the plants by means of recombinant DNA, plants might be produced that would have the independent ability to fix nitrogen. If this could be done with crop plants, the use of nitrogen-containing fertilizers would no longer be necessary.

Some plants are resistant to certain pests and diseases, and others are not. If the genes that are responsible for the resistance to each of these diseases could be identified and isolated, and transferred into non-resistant plants by means of recombinant DNA, it might be possible to convert susceptible plants into resistant plants.

The protein produced by some plants are deficient in one or more amino acids. For example, corn is deficient in lysine. Lysine is one of the eight amino acids which cannot be synthesized in the human body. Humans must, therefore, obtain this amino acid from their food. People whose diet does not include milk, eggs, meat, or other such sources of protein, but whose sole source of protein is corn, develop a deficiency disease known as kwashiokor, caused by an insufficient supply of the amino acid lysine in the diet. If the genes responsible for the synthesis of lysine could be transferred into plants which are deficient in lysine, these plants would then produce lysine and could be eaten as the sole source of protein without inducing a deficiency disease.

Several reports of the transfer of bacterial genes into plants using recombinant DNA techniques have appeared in scientific journals. A recent article (see E. C. Cocking, *Nature,* Vol. 266, March 3, 1977, p. 13), however, has suggested that the general view now is that the evidence for the modification of plant cells by bacteria DNA is circumstantial and weak. Much further work must be done before we will know whether this actually can be done.

Use of Bacterial Plasmids

One possible approach may be the utilization of the plasmids from *Agrobacterium tumefaciens,* bacteria which cause crown gall disease in certain plants. When this organism invades these plants, tumors are formed near the crown of the plant, that it, where the root meets the stem. It has been found that *A. tume-*

faciens contains plasmids (called **Ti** plasmids for tumor-inducing), which are transferred from the bacteria to the plant cells. It is apparently the presence of this **Ti** plasmid in the plant cells that transforms them into tumor cells.

In the same way that other bacterial plasmids have been used to carry foreign DNA into bacteria in recombinant DNA research, it is hoped that **Ti** plasmids may be used to carry foreign DNA into plants. If appropriate enzymes can be used to open the circular DNA of the **Ti** plasmid, attach the foreign DNA, and reform the new enlarged circular DNA of the plasmid, it might be possible to use this plasmid to carry the added foreign DNA into the plant.

There are some problems with this approach, however. In their present form the **Ti** plasmids produce tumors. Eating tumors may not appeal to some people, and such tumors certainly are not of benefit to the plant. There are possible ways to avoid this difficulty, however. Mutations in these plasmids are known which cause them to lose their tumor-inducing ability. Thus, one of these mutants could be used. It is also known that apparently normal plants can be generated from some kinds of plant tumor tissue and that these plants still carry the **Ti** plasmids.

Another problem is that the **Ti** plasmids are not found in the seeds produced by plants with crown gall disease. Thus, if these plasmids were used in recombinant DNA work, cuttings of the plant instead of seeds would have to be used for propagation. Furthermore, *A. tumefaciens* rarely infects monocotyledons (the seeds of monocotyledons produce a single leaf as they sprout, while the seeds of dicotyledons produce two leaves as they sprout). Some of the most important crop plants, such as corn and wheat, are monocotyledons.

No doubt many problems must be resolved before genetic modifications of plants by recombinant DNA techniques can succeed. We have no assurance at this time that ultimate success will be achieved. At least, as far as we know, plants do not have special organs or tissues similar to those of higher animals that endow them with the abillity to destroy foreign substances. Once it is possible to introduce foreign DNA into plants, therefore, it is not likely that this material will be destroyed. Whether it will be incorporated into the genetic apparatus of the plant and function to produce its normal product yet remains to be seen.

One Beneficial Result of Recombinant DNA Research

Whatever else may or may not be accomplished by recombinant DNA research, one very beneficial result of this research will be a great increase in our knowledge of genetics. Recombinant DNA research provides us with tools that will help us to understand the function of various genes and perhaps help us to understand the relationships that exist between these genes. Such legitimate potential benefits of genetic research should not be needlessly impeded because of objections to or fear of other aspects of this sort of research.

Chapter 18

Controlling Our Own Evolution? An Impossible Dream!

Now let us leave this more limited area of genetic research and consider the area of genetic engineering that is sometimes referred to as "controlling our own evolution." Will man ever be able to deliberately alter his genes in some purposeful preordained fashion such that he could actually improve his basic physical, mental, and emotional characteristics? We are not talking about merely correcting some defect in a gene that produces a genetic disease, such as sickle-cell anemia, but we are referring to a basic alteration in the genetic makeup of man.

In the first place, it is very unlikely that any change at all in the genetic apparatus of man will effect an improvement in a normal human being. We predict that any alteration in a normal, healthy gene—or any change in the normal way these genes are regulated in the normal, healthy human being—will produce only harmful results. Believing, as we do, that we are "fearfully and wonderfully made" (Psalm 139:14) by God our Creator, then normal man is as perfect as he can be, limited as he is by his human status. One might conceivably visualize a form of life superior to human life, but such could not be achieved by remodeling human genes or altering, somehow, the overall makeup of the human genetic apparatus. An automobile is not a remodeled horse and buggy. The automobile was an entirely new creation.

Attempting to remodel a normal human being would almost certainly have disastrous results, and it certainly would not cause him to evolve to a higher form of life.

Our basis for believing that attempts to remodel or improve a normal human being can have only disastrous results is well founded. If we intend to alter the genetics of the human race, these alterations must be achieved on the germ cells—the egg and sperm. We must somehow remove and chemically manipulate the genes of these cells and then replace them without causing injury. Presently, we have no technique for doing that—and may never have.

The nuclei of these cells contain at least one hundred thousand (and perhaps several million) genes. Right now we do not have the slightest idea which gene is which, and even if we did, we do not have any way of selectively removing any particular gene. If we did have some way of selectively removing a gene, we would not have the slightest idea what to do to the gene (assuming this was possible at all) to change the characteristic governed by the gene in some beneficial way.

An Immensely Complex Problem

As a matter of fact, it is simply inconceivable that this could ever be possible. We know that every characteristic is under the control of more than one gene. For example, eye color in the fruit fly, *Drosophila,* is under the control of as many as 15 genes. Furthermore, each gene affects more than one characteristic. This gives us at least some idea of how complex our genetic apparatus is.

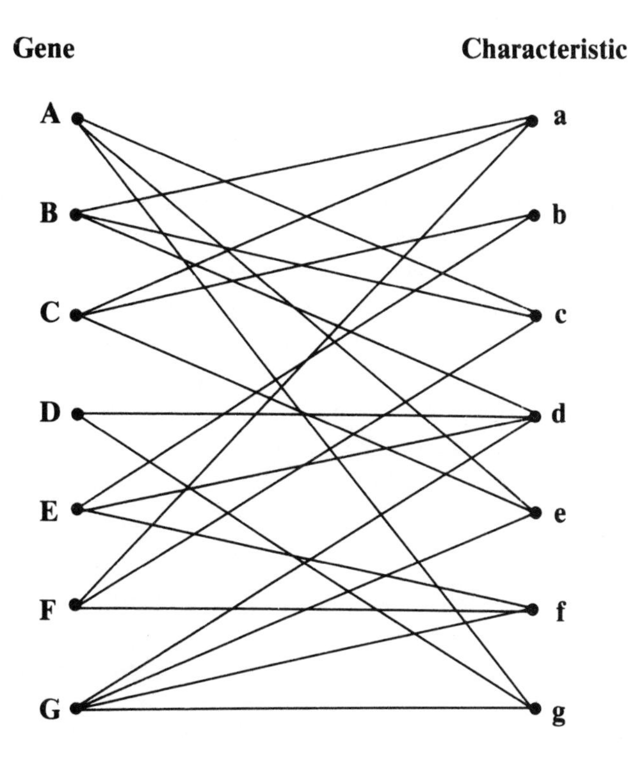

Figure 5. Hypothetical relationships between genes and characteristics.

Let us suppose now (see Figure 5) that we want to change gene **A** in order to cause some specific change in characteristic **c**, one of the characteristics affected by gene **A**. In the first place, as mentioned earlier, we do not have the slightest idea how to change a gene in order to bring about a certain desired effect. But supposing we *did* know how to change gene **A** in order to alter characteristic **c** in a certain way. Note that gene **A** also affects characteristics **e** and **g**. What is going to happen to characteristics **e** and **g** as a result of the change in gene **A**, which we have somehow engineered to alter characteristic **c**? Whatever happens, it is almost certain to be disastrous.

Note further that characteristic **c** is also controlled by genes **B** and **F**. To really change characteristic **c** in some preordained beneficial way, we would have to change genes **A**, **B**, and **F**. If that problem is not tough enough, note that genes **A**, **B**, and **F** control not only characteristic **c**, but also characteristics **a**, **d**, **f**, and **g**! As impossible as the problem might be of conceptualizing how we might change a single gene to change a single characteristic, we can now see how impossible the problem really is when considered from the overall viewpoint of our incredibly complex genetic apparatus.

Even if we *did* know how to do all of this, we have no way of specifically altering a gene without resynthesizing the entire gene. Genes consist of thousands of units. The projected alteration might require a change in one or several of these units, but we have no way of bringing about specific changes of this kind, nor is any such technique even conceivably possible. Thus, the entire gene would have to be synthesized, incorporating the desired changes. Even though such a synthesis is conceivably possible (and the synthesis of a very small gene has already been accomplished, although it took a team of scientists nine years!), it presents a practical impossibility.

We can thus see that the idea of "controlling human evolution" is an impossible dream. Just as ideas concerning the evolutionary *origin* of man are dreams of an impossible past, so is the dream of "controlling our own evolution" a dream of an impossible *future*.

Part IV

Artificial Insemination and Test-Tube Babies

Chapter 19

Artificial Insemination

Human cloning is still in the realm of science fiction, but "test-tube babies" are now fact. The door was opened by artificial insemination, the use of which is increasing rapidly.

In his article, "The Obsolescent Mother: A Scenario,"[1] Edward Grossman makes the point that artificial insemination is leading to an extra 20,000 births a year in the United States. Some 10 years ago a poll showed that only about 3% of Americans had even heard of artificial inseminaton. The facts now are that approximately 1% of all children born in the United States are the result of artificial insemination.

Banks are being established so that semen can be stored, and this could be used for the impregnation of millions of women, even many years after the death of the donor. The terms AIH, which means "artificial insemination by husband," and AID, which means "artificial insemination by donor," have become recognized as part of the terminology of modern medical science.

It is a fantastic fact that one human ejaculation

---

[1]*Atlantic*, May, 1971, p. 47.

contains something like three hundred million sperm, each of them being approximately five hundredths of an inch long. One single thimble could hold approximately three billion human sperm, sufficient to almost double the world's population.

Artificial Insemination of Cattle

In the animal world we find that about 95% of all the cattle born in the United States (this representing about 60 million head) are actually the result of artificial insemination. Most of this is accomplished by using sperm supplied by prize bulls.

This sort of insemination is also used on 50 million ewes, one million sows, and many other animals in the United States.[2]

One of the practical aspects of this work is that reproductive biologists have learned how cows can be caused to "superovulate" by the administration of certain hormones. This causes numerous eggs to mature prematurely, and then to be released from the ovaries of the cow. This is followed by artificial insemination of these cows. Eventually the fertilized ova are flushed through the fallopian tubes of these prize cows. They are then transplanted one by one into the uteri of other cows that were not so genetically valuable, but could satisfactorily produce the calf that was the union of the sperm and egg of a prize bull and a prize cow. By this process one prize cow could be made to produce hundreds of progeny, instead of seven or eight in a lifetime, and never actually go

---

[2]These figures are provided by Robert T. Francoeur in "We Can—We Must: Reflections On The Technological Imperative," *Theological Studies,* Vol. 33, No. 3, September, 1972, p. 431.

through pregnancy itself.

A cattle embryo has also been implanted in the uterus of a live rabbit, which serves as a satisfactory temporary host. They have then been taken out and successfully reimplanted in foster mothers, who would in turn give birth to normal calves.

Experiments with calves have also been undertaken, whereby the embryos of calves have been frozen at temperatures lower than 400 degrees below zero F, then thawed out and reimplanted in surrogate mothers. They have been brought to successful birth as normal calves.

Experiments With Humans

It was not long before experiments were taking place with humans, especially in cases where the woman's fallopian tubes were blocked. Her own egg would be taken and fertilized in a special type dish with sperm from her husband, and then replanted into her own womb. In other cases where a woman could not produce her own ova, the routine would involve another woman's fertilized egg that would be transplanted into the nonegg-producing woman who would then carry the child in a normal way.

The birth of the "test-tube baby," Louise Brown, has made it clear that at least one type of these experiments can be satisfactorily concluded. Now it is also true of a woman carrying a child on behalf of another woman, as above. We shall discuss certain ethical questions associated with the procedures, but these activities are clearly of a different category from genetic engineering where the DNA itself is altered. This is discussed more fully in Chapter 20.

There is possible truth to the argument that even if a so-called monster were developing by such techniques, it is likely that there would be a spontaneous abortion (miscarriage), for that is a common result with the human fetus when an abnormality has developed. In addition, much information about the fetus can be determined by means of prenatal diagnosis. Thus, the likelihood of the development of a monster would be known long before the child had reached term inside the mother's womb.

The increase in the number of human sperm banks we have referred to has been remarkable. Such banks are located in 12 cities in the United States, with four times as many cases in 1976 as there were in 1972. It is now becoming possible for a mother to decide whether she shall have a male or a female offspring.

Obviously, there are social implications. There have even been legal problems as to the legitimacy of children born as a result of AID. Something like one in every 200 Americans is now born as a result of this method, and some courts have ruled that such children are legitimate. Other states have declared that they are not legitimate, and in some states the woman who is impregnated by AID is considered guilty of adultery.

Law Suits Have Begun

Already there are ominous implications, such as law cases where parents have separated. Courts have given opposite decisions as to whether such a "father" has the right to visit, or indeed whether he has any rights at all. It would even be possible to argue that the anonymous donor would also have visiting rights in certain cases.

Yet another possibility—and one that has actually gone before the Courts, with judgment in the husband's favor—is that a woman who gave birth to a child by AID without her husband's knowledge or consent could be divorced on the grounds of adultery.[3]

---

[3]Reported by Howard & Rifkin in *Who Should Play God?*, p. 101.

Female Ova Banks, Too

The commerical aspects are not limited to male sperm, for female ova banks have also been established at various universities. These new developments touch the lives of us all, women as well as men. They have even intruded into the lives of our newborn offspring.

It is not generally known that approximately 90% of newborn babies in the United States are genetically screened before they leave the hospital, even though the parents do not always know of this procedure. The argument is sometimes advanced that charts will be developed whereby it will be known what the individual's genotype is at birth, and therefore the mating of couples having the same genetic defects will be prevented. We have already said that this overlooks the human factor, and even the likelihood of sharing because of a mutual problem.

Where does it all take us? Are we to be classified by scientists as mere machines, with numbers instead of names? Will we be categorized according to our genetic suitabilities? This possibility is being discussed!

Chapter 20

"Test-Tube Babies" Are Here

The whole question of "test-tube babies" with its attendant moral issues has come into new focus, with sensational headlines around the world. Dr. Patrick C. Steptoe, a British gynecologist and researcher, together with Cambridge University's physiologist Dr. Robert G. Edwards, have announced that the world's first "test-tube baby" was born on July 26, 1978. She was named Louise and was the child of Mr. Gilbert John Brown and his wife, Leslie. It had not been possible for her to conceive normally because of blocked fallopian tubes. The baby was 5 lbs. 12 oz. at birth, and a number of other details were given. The successful test-tube conception had taken place on November 10, 1977, when Dr. Steptoe obtained an unfertilized egg from the body of Mrs. Brown, and mixed it with her husband's sperm.

The supposed "test-tube" was not really a test-tube, but was a glass container which is known as a petri dish. To insure that at least one healthy fertilized egg is produced, the prospective mother is first treated with a hormone that causes "superovulation." That is, instead of an ovary producing a single mature ovum, as is usually the case each month, an ovary is induced to produce several mature ova. The eggs are removed by microsurgery and mixed with her husband's sperm in a petri dish. The eggs then each begin to develop by cell division.

A few day after the eggs were fertilized, they had developed sufficiently to be inserted into Mrs. Brown's uterus. One of the developing microembryos was selected for insertion into Mrs. Brown's uterus (the others were discarded). There it attached itself and obtained nourishment from the mother in the normal way. Thus her damaged fallopian tubes were bypassed, and the rest of the development followed more or less the same processes that are involved in a normal pregnancy.

Disposing of Embryonic Life

Yet another challenge came from J. F. ter Horst, writing in the *Los Angeles Times* of August 1, 1978. His article was entitled "Horde of Implications Born Along With Test-Tube Baby." He asks the question, "Does man have the right to induce human life in a glass dish? As of July 25, 1978, the moral, legal, and political imperatives of that question ceased being matter for future speculation—the future is now."

J. F. ter Horst is a syndicated columnist in Washington, and he was referring to the birth of Louise Brown.

Ter Horst quotes a number of authorities. He asks if the doctors are performing a new kind of abortion, because they are fertilizing several eggs from the prospective mother, and then discarding all but one after careful microscopic study. He makes the point that suddenly the implications are enormous, and so he asks, "Will the doctors be performing a new kind of abortion?" He quotes Dr. Andre Helligers, the gynecologist who directs the Kennedy Institute of Ethics at Georgetown University, as saying "It begins to make the child a pure consumer item." He asks who has the right to the other healthy tiny embryos—the parents? Do the doctors concerned have the right to dispose of the embryos, even though they consist of only a few cells? Do they have any rights in the rearing of the child?

As ter Horst points out, there is likely to be a new national controversy over abortion, centering on the issue of the right to life of a fetus. He asks, "Is an embryo that emerges from a laboratory dish any less human than one that results from normal conception?" He asks a further relevant question, "Or, to go back to Square One, is the laboratory process to be condemned as unnatural, tinkering with cosmic creation, and therefore to be rejected—by law, if necessary? Scientists, theologians, and politicians will be wrestling with this issue for years to come—starting now."

Ter Horst finishes with a very interesting statement. He comments on the normal mother's opinion that heaven is a mother with a baby in her arms. He then adds this warning: "But we should also remember one other thing. Heaven, by definition, is merely a step away from hell." As we have outlined in this presentation, there are very serious difficulties involved in this whole concept of test-tube babies, and even more difficult are the issues attached to the new concept of genetic engineering and recombinant DNA.

Concerning "test-tube babies," there are, of course, many arguments pro and con. On the one side it can be argued that every time a woman's ovum is fertilized by a human sperm, 23 chromosomes from each partner come together and an embryonic human life is procreated.

We have just seen that when this takes place outside the womb in laboratory conditions, there is a great deal of selectivity as to which embryo will be implanted back into the woman's womb. This necessarily means that some developing embryos are then disposed of. They are literally washed down the sink. Some would answer this by saying that, after all, this particular woman would never have produced a child unless this outside-the-womb conception had taken place. They further argue that if the husband of such a woman was true to his marriage vows, and accepted Christian principles of faithfulness to one woman, his sperm would never result in a child in any other way.

While there is some logic in this argument, the other side is that no matter how the sperm and egg come together, once they have united, there is embryonic human life. When the embryo is destroyed, someone is taking on himself the tremendous responsibility of disposing of that life, those embryonic infants that are not selected for implantation.

Another moral argument sometimes used in favor of the "test-tube baby" is that such a baby is very positively desired and planned for, often in a way that is very much more definite than with the child who may result from normal intercourse. The same is true even when the baby results from intercourse within the marital relationship, where it is often an act of physical gratification, with no premeditated desire for a child to result from that act.

Could a Test-Tube Baby Sue a Scientist?

Another article in the *Los Angeles Times* of August 7, 1978, made the front page, alongside the bold headlines proclaiming the death of Pope Paul VI. The article in question was headed, "Bar Considers Moral Issues of Test-Tube Birth," being written by Jim Mann, a *Times* staff writer. He asks, "Could a test-tube baby sue a scientific researcher on the grounds that his conception had been a mistake? What would happen if poor women began to agree, for a fee, to become surrogate mothers and bear children for wealthy families? What if couples ask scientists to separate sperm cells in an effort to make sure that the test-tube babies they were planning would be boys?"

Mann says these were just some of the questions and problems that were raised by a group of scientists, lawyers, and educators at the American Bar Association's Annual Convention in New York. He quotes Professor Ethan Signer of the Massachusetts Institute of Technology as stating that things were happening very, very fast, and that if we did not take notice and do something about it, it would be too late. Opinions were expressed that all the scientists and lawyers present believed that the public should make at least some of the fundamental decisions regarding genetic and molecular research.

The members of the convention were not sure that could be done. Professor Norton Zinder, Professor of Molecular Biology at Rockefeller University, is quoted as stating, "I do not believe that we have the forum necessary to solve these problems." He added that there was no way to have an analytic discussion with the legislators, because they did not seem to understand the problems of genetic research. Arguments were put forward to suggest that judges were likely to "err on the side of safety. They would halt something even indefinitely, even until a final judgment can be made."

Clearly the Convention was confused with various suggestions such as a Presidential commission being urged, or a special Congressional committee, with power to oversee biologic and genetic research. It was recognized that such a commission would eventually be dominated by scientists, anyway.

Even the scientists on the panel differed as to the hazards likely to be created by test-tube births and "cloning." There was also concern as to the intrusions likely at the level of politics. Professor Signer is quoted as saying, "Some people are going to get to determine what the rest of society looks like."

A Law Suit Following Destruction of an Embryo

Already there has been a court case relating to a so-called test-tube baby (in fact, an embryo). Dr. Landrum B. Shettles is a distinguished researcher associated with the Columbia Presbyterian Medical Center and the Columbia University College of Physicians and Surgeons in New York City. He claimed to have successfully brought together the sperm and egg of a Florida dentist and his wife. Shettles' superior officer at the Columbia Presbyterian Hospital, Dr. Vande Wiele, had confiscated the specimen and destroyed it. The entire attempt had been terminated because the higher official believed that Shettles was attempting to create what could be a monster. Eventually, Dr. Shettles resigned, and the Florida couple have initiated a $1.5 million law suit against Dr. Vande Wiele. They claim that his action in terminating the possibility of the embryo being implanted into the woman's uterus had prevented them from having a child. They claim, also, that it was their last opportunity to have such a child.

Dr. Shettles had been working in cooperation with Dr. M. J. Sweeney, a clinical professor of obstetrics and gynecology at Cornell University Medical College. It was claimed that the embryo was incubating, and would soon have been ready for implanting into the uterus of the dentist's wife. Her infertility was due to blocked fallopian tubes.

Surrogate Mothers

We referred above to the report by Jim Mann. One of his questions was, "What would happen if poor women began to agree, for a fee, to become surrogate mothers and bear children for wealthy families?" As this book goes to press, such a prospect is no longer mere speculation. Newspapers have widely reported the case of a healthy woman—already the mother of children—who was artificially inseminated with the sperm of a man who was not her husband—a man she did not know—bearing his child, then handing it over for a previously agreed fee so that the other couple could have their "own" child. The woman supposedly had noble intentions, and presumably she did. We make no suggestion otherwise in this particular case. She already had a sufficient family by her own and her husband's intentions, so she answered a newspaper advertisement to bring happiness to another couple. Her body became the incubating machine for his child.

All very noble, or so it was reported. However, let it be remembered that in this case it was also *her* child. The ovum was *not* that of the donor's wife. Clearly there are moral issues involved in giving up one's own child, in this case to an unknown, able-to-pay father.

Another major problem is that the floodgates are now opened. Moral principles are too easily thrown out the door when financial rewards are offered. Prostitutes sell their bodies for money; millions of unborn babies are destroyed each year, very often because of *lack* of finances. Bring the factors together, add the concept of surrogate motherhood, and a new social pattern is introduced.

Indeed, the prospect is even worse. Now it will be possible for both sperm and ovum to come from biological parents who simply don't want the inconvenience of nine months baby-bearing. So why not hire another woman's body? Indeed, many a teenager of twelve or thirteen would be very satisfactory. To fantasize, bring them in by the thousands from the poorer cultures of the world—select them carefully and feed them well, of course . . . our offspring must be healthy! (Maybe the fantasy is not so far-fetched as it *should* be . . . our science fiction has a strange way of becoming fact, whether we like it or not!) The prospects surely are frightening—and morally reprehensible.

This is only one of the ways in which dangerous precedents are being established in our frighteningly capable scientific generation. We shall come back to the subject of using human bodies in new ways.

"Test-tube babies" are now a fact of life. The birth of Louise Brown demonstrated that. Several similar births have already been announced.

Such babies involve fertilization of female ovum with a male sperm. The only part of the procedure that is unnatural is the fact that fertilization occurs outside the human body. The production of the sperm and ovum, and the nurturing of the tiny embryo all occur in the human body in the natural way. That is a very long way from the cloning of a human baby. In fact, as we have seen, it is an entirely different process. We have discussed it mainly because the two processes tend to be brought together in popular thinking.

Part V

Cloning, Creation, And The Golden Age

Chapter 21

A Divine Boundary Is Set

Does cloning transgress the will of God? What about recombinant DNA? Do the spiritual dangers outweigh the blessings? There are many such questions which we begin to consider in this chapter.

Are there any clues from the point of view of theologians? Indeed there are. While not all theologians accept the Biblical concept of creation, yet the teachings of Christianity must ultimately come from the Bible, and the Bible contains clear statements relative to creation. We shall confine ourselves to aspects of that subject that touch on the basic issues discussed in this book.

"After His Kind"

In the first chapter of Genesis, the first book of the Bible, we read of various acts of creation. At Genesis 1:11, we learn that the earth was to bring forth grass, and herb yielding seed, and fruit trees bearing fruit "after their kind, whose seed is in itself upon the earth." We find that there is a repetition of this concept of "after its kind," as with the creation of living creatures after their kind (verse 24).

When we come to the creation of man, we do not read "after his kind," but instead we read, "in our image, after our likeness." We read of God saying, "Let us make man in our image, after our likeness: and let them have dominion over the fish of the sea, and over the fowl of the air, and over the cattle, and over all the earth, and over every creeping thing that creepeth upon the earth" (Genesis 1:26).

We then read in the next verse that God created man in His own image, "in the image of God created He him; male and female created He them." Man is to be fruitful and multiply, to replenish the earth and to subdue it, to have dominion over the fish of the sea and over the fowl of the air, and over everything that moves upon the earth. Then at the end of the chapter we read that God saw everything that He had made, "and behold, it was very good."

God has confined the plants, the insects, the animals, and the fish "after their kind." Barriers have been set. It is not always possible today to know exactly where those barriers are. The classification of plants and animals into species, genera, families, etc., is often determined rather arbitrarily according to the opinion of the particular expert in a limited field of zoology, or whatever the particular science might be. Nevertheless it is recognized that there *are* boundaries which are set, and that interbreeding beyond those boundaries does not take place.

Crossing Boundaries With Mingled Seed

The Divine decree has been established to set those boundaries. It surely was never the Divine intention that experimentation should be undertaken with life forms beyond their "kind"—for example, between a mouse and a rabbit. It certainly was never the Divine intention that there should be a mingling of seed of a mouse and a man. As we have seen in this book, such experiments have indeed taken place, at least in cell culture.

Man is made in the image of God, quite distinct from all other species. Undoubtedly the Divine barrier has been set against any co-mingling with non-human species. Indeed, in other ways that were clearly understood by people of an earlier time, the Divine judgment was pronounced against practices associated with false religions, such as those in the city of Ur, and among the Canaanites. We read in Romans Chapter 1 that God gave up those who offended in serious religious practices. As we study these ancient people we find that there were various sexual involvements, such as sex between humans and animals, and the judgment of God is made clear in cases where forbidden barriers were crossed.

God Himself set the boundaries. Man was a separate creation from all other lesser beings. Bringing together a mouse and a man is a clear transgression of the divinely established order.

Eve Was Not Cloned!

As we go on in the record of Genesis Chapter 2, we read of the formation of Eve. There is no better way to explain the oneness of man and woman, and yet their distinctiveness as separate human forms. This was God's appointed way, His act of creation of man first of all, and then the creation of woman from that man. It was a woman from a man: beings of opposite sex. That is the opposite pattern from what we find in cloning.

Some have actually suggested that the possibility of cloning is included in Genesis 2:22, 23 where we have this record of God taking a part of man's body and making woman from that part. When she was brought to Adam, the man himself said, "This is now bone of my bone, and flesh of my flesh."

However, that is not cloning. It was God's way of forming a woman who would be one with man and yet different from him. Cloning is quite different, in that all the chromosomes of one being have been brought together in the one body produced from the clone. The sex must therefore be the sex of the donor cell. There could be only one immediate parent, and that parent would be man or woman, according to which donor cell was taken.

In the formation of Eve, God was demonstrating the oneness of man and woman, and yet their total separateness. There has never been a better explanation for the original formation of man and woman as different and yet so intimately related, needing each other for procreation. Eve was female, not male as was Adam. This is diametrically opposite to what would have been produced by cloning.

Obviously cloning crosses another Divine boundary at this point. God ordained marriage between a man and a woman, and the divine instruction was that a man would leave his father and his mother, be joined with his wife, and the two would become one flesh. With cloning, the need for such a union is eliminated. The man himself, or the woman herself, would simply contribute a tiny cell, and after "processing," progress would continue in a similar way to natural development. The embryo would make use of a female body, merely as a necessary hatchery in which the cells could divide and multiply.

As we have seen, if the person donating the cell nucleus was a woman, she could even have that development take place in her own body. She could totally do away with the need for any man, or any other person other than herself, so far as providing the nucleus, the enucleated egg cell, and the actual development were concerned. This is not God's method of procreation, and we shall see that it is likely to involve Divine judgments.

Man is Prevented from Partaking Of the Tree of Life

As we go on in that Genesis record, we quickly come to the story of the fall. Man ate the forbidden fruit, gained knowledge that he was not meant to gain, and was put out of the Garden of Eden. We read at Genesis 3:22 that the Lord God said that the man had become "like one of us, knowing good and evil." God decreed that man should be put out of that place of blessing, "lest he stretch out his hand and take also from the tree of life, and eat, and live forever."

As a result of this decision, man was sent forth from the Garden and made to cultivate the ground by the sweat of his brow. The Divine restriction was imposed against Adam and Eve taking of that Tree of Life which would allow them to live forever. Spiritually, a Christian has partaken of that Tree of Life, for Jesus Christ has come and given him eternal life. By dying on the cross He has put to flight the forces of evil, making it possible for man to know eternal life. He has declared, "I am come that they might have life, and that they might have it more abundantly" (John 10:10). "I am the way, the truth, and the life, no man cometh unto the father but by Me" (John 14:6).

However, our first parents could not eat of that tree —not even of its leaves. They were put out of that Garden, for their own sakes. How dreadful it would have been to live on forever, in bodies having increasing pain from disease, and so much more. God intervened, and drove them out. The Divine barrier was set. Man would dare to intrude at the risk of death. That is made clear by the presence of that being with the flaming sword . . . and the barrier has never yet been lifted. Man ventures at his peril.

Chapter 22

God Intervenes In His Own Time

God had ordained that one man and one woman would be the procreators of all human life that was to follow. He ordained monogamous marriage as the institution for procreation, and Eve is called "the mother of all living" (Genesis 3:20). It certainly is a travesty to suggest that it would be within the will of God for life forms to be developed utilizing creatures that were less than man or woman.

So we read in that chapter that man was driven out, in case he attempted to partake of the Tree of Life. There is a very real sense in which man is at least beginning to partake of the Tree of Life, with the present activities involving genetic engineering. Many may well say that he can do as he pleases, but ultimately he cannot. It is still true that God is on the throne as Ruler of all. He is still in control of all life's happenings.

Are we suggesting that God will intervene? The answer is that in His time He will. He might very well be saying in effect right now, "Thus far shalt thou go, and no further."

A Warning From the Tower of Babel

Let us go on to another illustration from that Book

of Genesis. At Chapter 11 we read of the incident of the Tower of Babel. Man was making his tower to the gods of heaven, for that is believed by many to be the concept of the top being "unto heaven." That term "heaven" is an abbreviated way of saying, "The gods of the heavens." The one true God was displeased, and this is what we read, beginning at verse 6:

> This they begin to do: and now nothing will be restrained from them, which they have imagined to do.
>
> Go to, let us go down, and there confound their language, that they might not understand one another's speech.
>
> So the Lord scattered them abroad from thence upon the face of all the earth: and they left off to build the city (verse 6-8).

The people went too far. They were actually building that city and that tower. It seemed that they were "getting away with it," totally disregarding the one True God Who demanded holiness and separateness of worship toward Himself.

However, they did not get away with it, and at the appropriate time judgment fell. Judgment does not always fall immediately, for God is merciful and patient. That is demonstrated in Chapter 6 of Genesis where we have the story of the Flood. At verse 5 we read of God seeing that the wickedness of man was very great, and that his every imagination was evil continually. God determined to send the judgment of the flood, but one man and his family found grace in the eyes of the Lord. Ultimately that family escaped in the ark they prepared in obdience to God's instructions. It was not the next day after men rejected God, or even the next year, but a long period of time passed before that judgment fell. God is not limited to time as we are. In His own time He will protect His rights in regard to the Tree of Life.

It is at least possible that we are again facing a situation like that of the Tower of Babel. Our point is that God was displeased with man's intrusion into divine prerogatives, and possibly we are witnessing another intrusion. It is one thing to unite a male sperm and a female ovum outside the body, even though there are serious problems with that, as we shall see. It is quite a different matter for man to reproduce a human body from one cell. It has moral and ethical overtones that have very serious implications. Any one involved in such experimentation is possibly asking for the judgment of God on his actions. Such a statement is not written lightly, but is simply a statement relating to spiritual principles that are laid down in the greatest Guidebook of all times, the Bible. God intervenes when He is ready, and there are many evidences of that intervention in the Bible.

Man insists on his own will, in both ancient and modern times. He paid the penalty in the Garden of Eden, again at the time of the Flood, and then at the Tower of Babel. Possibly he will again pay the price in this matter of genetic engineering. Scientists themselves know the dangers, and yet as we have seen, they have deliberately chosen to persist with their dangerous defiance. Somehow that is typical of our age, so often characterized by increasing license and lawlessness. In II Thessalonians 2, we find that such lawlessness is to be expected as the world rushes on to its own destruction.

There are many indications around us that we are approaching the great climax of the ages. Very possibly we are in the age where Christ will return for His Church. Those who hold to so-called futurist prophecy believe this to be the case, but, whether that be so or not, there are many indications around us that the world is racing toward its own destruction.[1] It is seek-

---

[1]Refer to *The Decade of Shock,* Master Books, P. O. Box 15666, San Diego, CA 92115.

ing to destroy itself with the possibility of nuclear war and other means.

Tasting the Leaves

Possibly another indication that man's time is running out is this whole matter of genetic engineering. Man has put forth his hand toward the Tree of Life. It is as though he has tasted the leaves, and he has the fruit in his hand, but he has not yet eaten it. It is not that he has not desired to do so. It is merely that he has not yet perfected the techniques.

Man has also basically said "No" to the Christ of God, the One Who can bring deliverance, redemption, peace, and purpose. Man has sown the wind, and he will reap the whirlwind. Coming events are casting their shadows, and it is entirely possible that the facts of genetic engineering are yet another pointer to the soon return of Jesus Christ.

That event is to be followed by an even greater period of lawlessness, which many believe will last for seven years. It will be a time when people experience great tribulation and the world submits to the anti-Christ whose rule will be of iron—ruthless and tyrannical. Only those who publicly display the mark of the beast will be able to buy or sell—rather reminding us of the suggestion given earlier in this book that genetic identification be marked on the foreheads of all newborn babies.

God does not intend man to be a race of superbeings. He Himself is the Author of creation. Genetic engineering might well be in the Divine prerogatives, so that man can benefit from his knowledge of the structure of the life which God Himself has created, but it is yet true that limits have been set. Undoubtedly man is transgressing those boundaries when he begins to mix human seed with the genes and components of life associated with mice, rabbits, and other animals.

We saw that Adam ate of one tree that was forbidden, and God clearly stated that the ground would be cursed for his sake (Genesis 3:17). God drove man out so that he could not partake of that other tree, the Tree of Life. Man's meddling and foolishness is known to God, and in God's time judgment must fall.

Overcomers Will Eat of the Tree of Life

Ultimately human life has a spiritual content. We read at Revelation 2:7, "To him that overcometh will I give to eat of the tree of life, which is in the midst of the paradise of God." We partake of the Tree of Life through coming to Him Who said, "I am the Way, the Truth, and the Life: no man cometh unto the Father but by Me" (John 14:6). We have the right to partake of that life—eternal life.

God never intended that death would be the ultimate for man, and in His wonderful plan of salvation it is possible for us to partake of that tree of eternal life. Man has within him the potential to know God, to come to God, to have fellowship with Him. This is because man is made in the image of God, and is a truly spiritual being—in a way that is not the privilege of any other created being. Man is unique in his capacity to enter into a spiritual relationship with God.

Any playing with the potential of life that is found in humans is an insult to God. Man is taking that life which is described as being in the image of God and relegating it to something very much inferior. Man is putting his intellect on the throne. He is doing what we read of in Romans 1:23 where they "changed the glory of the incorruptible God into an image made like to corruptible man, and to birds, and to four-footed beasts, and to creeping things." It is no wonder that we read three times in that chapter of God giving those people up.

Man Foolishly Challenges the Creator

Perhaps recombinant DNA will come into its own. Perhaps there will be cloning. So far, God has said "No" to the partaking, eating of the Tree of Life. Man has said "Yes." Man has attempted to step into the very throne of God, taking to himself the role of Creator. God has clearly said that He will not give His glory to another. The creation of life is the Prerogative of God. However, one of God's ways of testing and bringing men to a full understanding of spiritual realities is to allow the forces of evil to exert their influence. Testing is necessary so that man can deliberately choose to know God and to worship Him.

God Allows Testing

Man is given the privilege of being tested for just the short span of his life, and the Christian's certain hope is that of eternity to follow. The wise will hear and understand. He will recognize that there is an intensification of evil around us in many ways. Yet another of those intensifications is in the realm of power, as the need for God is laughed at and put to one side, with foolish statements about billions of years of evolution. Suddenly we are supposed to believe that man can "control his own evolution" and that this can be condensed into one generation—condensed in fact, into something closer to a decade than a generation.

Man thinks he will create in a decade now . . . not needing millions of years after all. He talks blithely of his creative capacity! Man is blind to the greatness of the Almighty God Who is still the Omnipotent and only Creator.

The Christian sees another side. He recognizes an intensification of Satanic activity, an increase of lawlessness, a preparedness to accept a delusion and deception rather than truth. The Christian recognizes all these as signs to which the writers of the New Testament pointed. The wise man will heed, and will look for life, not in genetic engineering, but in Jesus Christ.

Chapter 23

More Theological And Other Questions

In cloning, according to the popular idea, a man's self would live on in that new body. Then that could be perpetuated again, and again, and again. According to this hypothesis, there could be a successful series of bodies that would stretch through the centuries and even through the millennia. It is at least possible that God has said "No" to such a concept.

This is not to say that cloning will not take place. Indeed, we have said above that it is "possible" that God has said, "No." This also means it is possible that God will NOT intervene before cloning is an established fact. However, even if cloning of a human does take place, it is not really the same individual who lives on. Man is body, soul, and spirit, and no matter how many clones a man might have, they all would still be separate beings.

Each Clone Would Be Responsible Before God

Let us imagine for a moment that cloning becomes successful, and that a man did indeed have several clones brought into being. Each of them would one day look more or less the same as that man looks now. However, obviously they are not that man. He is already alive, responsible to God for his actions, needing to preserve his own body against sickness, to see that he is properly fed, and all the rest. Each clone would have its own individual responsibility, its own soul. To that extent it is quite clear that each clone would be a separate being. Once the man himself died, he would be buried, and he would not personally live on, no matter how many clones had been formed from his body. The molecular structure of his body would even become part of various other forms of life, because man is dust, and to dust he returns. The structure that was once his flesh will soon be available for other forms of life on the earth. This is not reincarnation, but is simply a statement of fact—that our molecular structures will be reutilized after we ourselves have been put to one side.

We are saying, therefore, that no clone would actually be a continuing, living form of a particular person. This is true even though it had virtually the same bodily characteristics of that person. This is true because only the genetic information that was present in that person's body had been duplicated in producing the clone.

We saw that it is sometimes said that we have already living clones in our midst, as in the case of identical twins. This similarity exists because identical twins come originally from one cell that divided into two cells which then separated and each developed into separate beings. Thereby it was possible for there to be not one body, but two.

However, this is not really the same as with cloning. With identical twins, there was the act of sexual intercourse or some artificial way whereby the female ovum and the male sperm were brought together in the act of procreation. Cloning involves only the one cell from the one body, and is not the combination of two bodies having opposite sex characteristics.

Would a Clone be Truly Human With a Soul?

Many theologians and scientists believe that the life of the new baby is there at conception, at that moment when the male sperm and the female ovum unite. From that point on there is a life that is the direct result of a union between a male and a female parent. Some would therefore ask if a clone would have the human spirit that is a distinguishing mark from lesser creatures?

Today the term humanoid is used in science fiction, as in the stories of beings coming from other planets. The word implies that these beings are like humans, but are not true humans. If cloning takes place successfully, as seems possible, will that new life be truly human?

The question has even been raised as to whether a human clone, if ever successful, would in fact have a soul. Even among Bible-believing Christians there are serious differences of opinion as to when life takes place. We have seen that most conservative theologians would hold that life is there as soon as the sperm and ovum unite. At that stage 23 chromosomes from the woman and 23 chromosomes from the man have merged to form a human life. All the potential for life is in that fertilized cell. Others claim that life is there at the time of dividing, and others will insist that it is slightly further along, when the life form attaches itself to the woman's uterus. There are other points of view, leading down to the actual time of birth and the taking of the first breath.

One interesting point is that when identical twins are born, those little lives come from only one sperm and the one ovum. However, no one would say that there is only one soul shared between the two human beings, and in the same way it simply would not be possible to insist that a clone did not have a soul in the fullest sense. After all, every human cell is unique, different from the billions of others in even the same body. This would be especially true in human cloning where a nucleus with 46 chromosomes would be placed in an enucleated ovum.

Opposite points of view to this can be taken. Cells multiply thousands of times per hour within our own body (as tissue replacement takes place), and nobody would argue that such replacement cells contain souls. On the other hand, if all the potential of a human being is in an embryo that potentially has the power to reason, it would be very difficult to say that such a procreation . . . or clone . . . or whatever it is called . . . would not in fact be fully human.

What do we say then? Would a clone be truly human? The answer is that he would indeed be human, for its life came from human life even though in a manner different than is usually the case.

Problems of Inbreeding

Certainly there would be all sorts of problems in the matter of inbreeding. In the wisdom of God there are tremendous possibilities of diversity and individualization, all resulting from the Divinely ordained way of sexual reproduction. That union between a man and a woman is God's ordained way whereby conception can take place.

The gene pool available from our multitude of ancestors has fantastic possibilities and combinations. Such a gene pool would not be utilized where the newly formed living being was simply a product of one cell whose structure was totally decided before it was ever planted in the womb of a chosen woman and the process of development took place. This sharply contrasts with the genetic mixing that takes place during the normal production of spermatazoa and ova and subsequent union of sperm and ovum that takes place at conception. That interaction simply would not take place with cloning, although the clone itself would be no more or less "fit" than the donor, since they carry the *same* 46 chromosomes.

All human cells contain defective genes, and a clone would carry exactly the same defective genes contained in the cell whose nucleus was used to produce the clone. Furthermore, the cell used would have been a body cell (for example, a muscle cell). Mutations in body cells are far more common than in the much better-protected germ cells. The idea that a superior human being could be perpetuated through cloning fails to take this very important consideration into account.

The Planned Modification of Man

One thought-provoking article by a theologian about cloning is that by Dr. Harold B. Kuhn, of Asbury Theological Seminary at Wilmore, Kentucky. His article, "The Prospect of Carbon Copy Humans," was published in *Christianity Today,* dated April 9, 1971. He points out that experimental genetics has now been linked with biochemistry, and that this involves a movement for the planned modification of man. He states his view that the possibilities for successful cloning are there, essentials being a willing prospective mother, a medical biologist, a gynecologist, and a bit of donor tissue.

Dr. Kuhn told of attending a course dealing with medical ethics at the Harvard Medical School. There were expressions of great concern about the possibility of human cloning, and other forms of parahuman reproduction. It was recognized that the resulting creature could well be defective, and also that the scientific world has not always been guided by ethical considerations. While purely technical problems might mean that the cloning process might be delayed for some time, it is certainly high time, Kuhn suggests, to decide a public policy in this whole matter, rather than allowing mere technical development to have free rein.

A "Holy Mystery" is Becoming Clinical

Dr. Kuhn raises the point that, although genetic engineering should not necessarily be opposed, there are serious questions, such as the degree to which nature would be contravened by clonal reproduction. He points to the humanistic overtones of such procedures, and suggests that "a holy mystery," linked to God the Creator Himself, is to some extent now being made clinical and largely impersonal. He mentions also, as we have done, that there would be a lessening of the recombination and enrichment of the human gene pool that takes place through the normal genetic mix. This "mix" would be by-passed, and the results would not be desirable.

Dr. Kuhn draws a parallel with other forms of pollution. Man has managed to pollute the environment, damaging his own world, and he is now on a path leading to the possible denigration of his own inner genetic environment. The birth of a child should be a matter of joy to a couple who believe that their union is of God. God did not provide them with a series of test tubes, but with sexual organs. The sex act was meant by God to be the source of reproduction and a source of pleasure rather than the basis for mere technological manipulation.

Cherished Christian Beliefs Put Aside

It is not too much to say that the whole concept of cloning is opposed to certain Biblical principles. This is relevant with relation to the true nature of sexual union, the role of parents, the place of the family, and much more that is fundamental to our way of life. Nevertheless, the evidence indicates that experimentation will continue. Science has no history of heeding ethical warnings, and there is no reason to think that such warnings will be listened to in this case.

Martin Ebon quotes Dr. Charles Stinson of Dartmouth College in Hanover, New Hampshire. His article, "Theology and the Baron Frankenstein: Cloning and Beyond," appeared in *The Christian Century*, of January 19, 1972. Dr. Stinson is not alarmed by the potential of cloning, and he appears to think that society will do much better in a "next time" that is just around the corner. However, such optimism is unjustified. Anybody who stops and looks back will recognize that history strongly indicates that today's advances in science are likely to be spoiled by tomorrow's degenerations in relation to nature and human welfare.

We saw that scientists themselves are all too aware of the Frankenstein-type monsters that may possibly be developed, and this is no longer just something for tomorrow. The first steps have already been taken. We are firmly marching down a path of foolishness, and it is likely to lead eventually to disaster. The frightening apocalyptic pictures of the Revelation, the last book of the Bible, might well involve some of the dramatic changes associated with genetic engineering. It is frightening to realize that a Pandora's Box may be opened, and the possibilities go beyond cloning.

Man is not ascending, but is descending. That is the story of human history. We *learn* more, but with it comes all sorts of negative aspects. The very fact that many of our genes are defective is a pointer to this fact. Mankind is all too capable of turning the benefits of scientific progress into instruments of horrific destruction.

Mystery Replaced by Mastery?

Lovemaking and the attendant act of sexual intercourse are divinely ordained as part of marital life. The result is the wonderful mystery of life, of new life in a crying baby who is the delight of parents, and grandparents, and so many others. Cloning sets out to do away with the mystery and replace it by the mastery of man. Mystery is replaced by mastery. We have seen that in a sense man is putting himself above God. He sees himself as the new creator, no longer having need of God Who has promised eternal life. Man wants life to be available to him by his own mastery of his body cells. It is, of course, a poor substitute. As is so often the case, man is being deceived into accepting a counterfeit. In his desire to live on, he attempts to bypass God through human effort, ignoring the fact that eternal life comes only from God through Jesus Christ.

In actual fact man is not doing what God has done. God has given life out of *nothing*. Man is *manipulating* those life forms, but he is necessarily restricted to the patterns of life that are already available. Man might recombine, manipulate, extend life, do away with some illnesses and diseases, but this is only possible because a pattern is already there. How can any scientist who thinks rationally about it be so foolish as to believe that the intricate pattern associated with DNA could have arisen by the laws of chemistry and physics, plus time and chance? The complexities are of such a nature that any unbiased person must acknowledge that behind DNA, one of the essentials of life, is the great Divine Builder Himself.

The fact is, cloning could lead to a new approach to human relationships. Human reproduction would become dramatically different from our present understanding of human love relationships. It would be dehumanized, and the present concept of biological parenthood would ultimately be replaced. If this exercise is taken to its ultimate, there would be no need for marriage involving a second person. A partner would be unnecessary. Obviously, before very long the family would be looked upon as a mere biological entity.

It is unlikely that human life in its wider meaning would be improved by cloning. Man would be exercising the right to make decisions in areas that have been kept from him, and he would thereby be intruding into the realms of God. He would be reaching toward that Tree of Life which is forbidden for man to partake.

It is one thing for mankind to want to improve the nature of human life, and the Bible makes it clear that medical aids are in order. Common sense tells us that it is perfectly in order to take legitimate steps to improve our health and welfare. It is quite a different matter to eat of the Tree of Life in a way that takes a massive step beyond procreation, toward what man himself regards as pure creation itself.

Man has always had some sense of responsibility toward the next generation. Many of those who discuss cloning suggest that there should be animal studies and experiments to bring the processes to perfection before humans undergo experimentation. Most would go so far as to say that all individuals concerned should be consulted, to ensure that the risks are understood. Such sweeping generalizations tend to ignore the rights of those new individuals who are to be objects of constant examination, analysis, and comparison. They are prime candidates to be children of confusion.

Cloning of humans, if it succeeds at all, is certain to produce many defectives, perhaps many more defectives than normal individuals. Either such defectives would have to be identified during development and then deliberately aborted or allowed to be born. Either alternative is repugnant.

Cloning and Man's Search For Immortality

We have said that man has a sense of responsibility toward the next generation. Allied to that is man's search for immortality. Right through the centuries men have talked about immortality and have sought for its secrets. Some believe that cloning is a partial answer to this search. It includes pride, for it touches man's own inherent demand for his unique personality to be perpetuated and even for his body to be preserved. With cloning, however, that new life will not be a continuation of either his body or his consciousness.

Every human cell is itself individual, and there is no guarantee whatever that the clone would bear more than a distinct resemblance to the donor. There would be many variables: the body of the woman who brings the cell to the mature life of babyhood, and then the environment in which the child is reared, together with the food and exercise provided over the years, the educational environment, the military demands, religious challenges, and so much more. This all makes it quite clear that there is no certainty whatever that a particular cell will ever mature to be an exact replica of the original donor. In any case, no matter how strong the resemblance, the clone would still be a totally separate individual. A person with five clones would still have his own separate body, with its own distinct personality.

Cloning is no more relevant for personal immortality than is the case with the continuation of one's personal name through a son born in the normal way.

However, there is a brighter side. We have considered many bizarre aspects of cloning, and we have emphasized the dangers besetting the use of this technique. That is the negative approach, and we do not withdraw anything that we have written.

What then is the brighter side? It involves theological concepts, and we outline some of them in our next chapter.

Chapter 24

The Golden Age Is Coming

Ever since Paradise was lost, man has been hoping for the golden age. Cloning and recombinant DNA have shown that such an age is at last a concept some take seriously, as something involving physical as well as spiritual characteristics.

Leaves . . . For the Healing Of the Nations

It is remarkable that the Bible gives a number of clues which possibly have their fulfillment in these new biological discoveries. Let us illustrate from Revelation 22:2 where we read of the Tree of Life which bore twelve fruits, and it yielded her fruit every month. Then we read this: "And the leaves of the tree were for the healing of the nations." The next verse tells us that "there shall be no more curse."

The curse on man commenced in the Garden of Eden. Man ate of a forbidden tree, and with his wife he was put outside that paradise. Sickness, disease, and physical death were part of the punishment. Every baby born since then partook of Adam's fallen nature—morally, spiritually, and physically as well. As the world's population grew, families gave place to tribes, and tribes to nations. Every one of those individuals, those families, those tribes, those nations, was contaminated by that original sin. The curse had spread like a cancer.

It is still spreading and will continue to do so until God intervenes. When He does so intervene, the golden age will be with us. That age must come, for otherwise God's purposes will have been thwarted. In all things He must have the preeminence (Colossians 1:18), and even the fall of man will be seen as bringing glory to God. Through it God called out a people to serve, love, and please Him.

There is to be a time of restitution of all things (Acts 3:21); there is to be a new heaven and a new earth (Revelation 21:1); we saw that the nations are to be "healed" (Revelation 22:2); and that there will be no more curse (Revelation 22:3). We also read that men will be able to eat of that once-forbidden Tree of Life (Revelation 2:7).

God did not create man as only a spiritual being, nor did He provide a garden to float in a nonmaterial heaven. Man has a physical body, and his home is solid earth. When everything is restored, it is reasonable to believe that man will again be physical, though not limited as at present; and his home will consist of material not unlike that of the present planet earth.

More Than Three Score Years and Ten

Furthermore, God did not intend that original man should live for only three score years and ten (70), or even for the 969 years that Methuselah lived. No—there was to be no time limit, for God Himself is eternal, and man was created to be a companion for God.

Ultimately man himself will be eternal—a physical, yet spiritual man, with an immortal and incorruptible body (I Corinthians 15:53).

The nations will be healed by the leaves of the Tree of Life. That will include all the benefits of a human genetic system restored to its original perfection such that all sickness, disease, and deformity will be eliminated. God Himself will be in control, and no scientific aberrations would be allowed.

There would be no more Down's syndrome babies, no more sickle-cell anemia, no more hemophilia, no more leprosy, no more leukemia, no more cancer. All those things would be done away with. Old age would be a thing of the past, and worn out limbs would be unknown.

No longer will the Tree of Life be out of bounds to man. The Tree of Life will be readily available.

Abundant Life in Christ

There is, however, one condition. It is available only to those who "overcome" (Revelation 2:7). Elsewhere we read that "overcomers" do so "by the blood of the Lamb" (Revelation 12:11). Access to the Tree of Life is available to those who know Christ as Savior, covered by His blood.

Possibly this is somehow tied up with the concept of the abundant life that Jesus offered (John 10:10). "Eternal life" is not only something extending beyond time: it is also the perfect quality of life.

In resurrection Jesus was still in a physical body, but not restricted by physical laws as He had been before the crucifixion. It was a new form of physical life. He could ascend into the heavens without any vehicle (Luke 24:51; Acts 1:9) and could walk through a closed door (John 20:19 ff). These activities go beyond the possibilities of recombinant DNA, but that is not surprising. Genetic engineering may be within the powers of human scientists, but God is not limited as they are, and there will be revelations and developments beyond human thinking.

Jesus could walk from Jerusalem to Emmaus, a distance of several miles, on nail-pierced feet (Luke

24:13 ff). He could prepare a meal for His disciples, despite the wounds in His wrists (John 2:9 ff). This was, of course, the miraculous work of God. Humanly speaking, physical healing was necessary for Jesus to be able to do those things.

No More Wars . . . Recessions . . . Or Social Barriers

A golden age is coming. There will be no armies, swords will be beaten into plow shares. There will be no competing political parties, for the Son of God will be the all-wise Ruler. There will be no religious groups or denominations, for the Head of the Church will be Christ Himself.

There will be no economic recessions or industrial upheavals, for the One Who taught the equality of men and absolute justice will be the Director of Industrial Relations.

There will be no educational imbalance, for the Master Teacher will be in charge of all schooling, no longer rejected by men, but able to communicate with an authority that none will gainsay.

Social and cultural barriers will no longer be relevant, broken down by Him who broke down a greater barrier, that middle wall of partition between God and man (Ephesians 2:14). Men will commune in one language, for the effects of Babel will be gone forever.

There will no longer be wild animals or other creatures, for the wolf will lie down with the lamb, the eagle will fly with the swallow, and the snake will share the hole of the rabbit. They will each know the voice of their Master, even the Creator Who will Himself be Lord of the earth.

Earthquakes will be nonexistent—likewise floods, tornadoes, and hurricanes. The elements will be under the total control of Him Who once rebuked the howling wind and the tempestuous waves so that they were immediately calmed.

Health Problems a Thing of the Past

Health problems will also be nonexistent, for the leaves of the Tree of Life will be available to all men then living, administered by the loving hands of the Great Physician. Genetic engineering will be a thing of the past, for when that which is perfect is come, then man's faltering second-best will be put aside.

The Golden Age is coming. ''Even so, come, Lord Jesus'' (Revelation 22:20).

Part VI

Abortion

Our purpose is not to investigate fully the subject of abortion, for it is a very complex issue. Nevertheless, a book that deals with cloning and DNA manipulation necessarily touches on "test-tube" babies. Once that subject is opened, we are close to the subject of abortion also. Hence the following outline is relevant.

We recognize that we will give little more than a summary, and we stress that there are already very good books and pamphlets available on the subject. Some of the statistics and relevant information in the pages that follow come from two of these books, the first by a medical doctor and his wife, and the second by a trained Christian counsellor. The two books are: (1) *Handbook On Abortion,* by Dr. and Mrs. J. C. Wilke, published in 1972 by Hiltz Publishing Co., 6304 Hamilton Avenue, Cincinnati, Ohio 45224 (Revised Edition, 1975); and (2) *Abortion, The Bible, and the Christian,* by Dr. Donald P. Shoemaker, published in 1976 by Baker Book House, Grand Rapids, Michigan.

What About Birth Control?

As we are at least briefly dealing with the subject of abortion, some would consider that various other ethical matters should also be examined. Obviously, the potential subjects could range far and wide, especially in the areas of sex relationships. If we deal with abortion, why not with birth control also?

However, birth control is not directly relevant to the subject matter of this book. It is actually an entirely different matter which we are not discussing. It touches on religious teachings and social ethics, but the use of contraceptives is not a means of taking life. Other developing methods, whereby the newly developed cell is immediately aborted, is another matter, and the arguments against abortion apply.

Contraceptives are justified by many as a means of allowing planned parenthood, this necessarily being the personal decisions of the couple involved. It is a way of postponing the commencement of a new life, rather than the destruction of that new life. It is not actually destroying life, for the sperm and the ovum are prevented from uniting.

Birth control involves ethical concepts as to the whole purpose of sexual relations and procreation, but it does not involve the decision to terminate a human life, as in abortion. We recognize the ethical and spiritual considerations. However, it is at least a totally different matter to practice contraception, with the idea of planned parenthood, than to practice abortion, which is an act of planned destruction of life.

Sterilization Is Not Abortion

Some would argue that sterilization is a related concept. While we are not endorsing sterilization, at least that is something done to the person's own body. This is especially true where it is an actual decision made by the person concerned. Compulsory sterilization for mentally defective people is another issue that we are not considering in this book. However, even where that is practiced it is an entirely different thing from the taking of a newly developing life which has been offered its temporary home by someone who now decides to expel that new life by killing it.

Chapter 25

Who Has the Right To Decide?

Some people claim that the mother has the right to decide the fate of the growing organism within her body. That totally overlooks certain basic facts. It is a human life within her, a separate human being in the fullest sense. Once the egg and sperm have come together at the act of conception, the mother is carrying another being.

The mother herself does not produce either the placenta or the umbilical cord by which her baby is atached to her. She does not even originally have the amniotic sac in which her baby is destined to live for that first nine months of its life. These three "life supports" all come from the cells within her uterus that are themselves forming a fetus. A cell was brought into being when the mother's ovum was fertilized by a male sperm. That cell contained within it the blueprint of those three systems, and so much more. Right from the early stages there are triggering points of maturation.

Maturational points are reached sequentially as a young child develops, and that is just as true with the stages of embryonic development. It is an amazing fact, for example, that the tiny human being has the inherent capacity to develop those three essential supports to life, supports that will be discarded once the time arrives for the baby to leave its mother's womb.

Economic Values and Abortion

Modern society tends to put an economic value on that developing baby, and sometimes even a physical value. If it is inconvenient for the mother to carry the baby because of her economic conditions, there is likely to be a sympathetic attitude toward her having what some refer to as "a little operation." If the baby is likely to be less than "perfect," then that is recognized by many as being a quite sufficient reason to terminate the life of that total human, though in embryo form, within the mother's body.

Such a concept overlooks the fact that literally all of us have imperfections. Our genetic structure is such that there is not a single human being alive who is perfect in the true sense of the word. Who shall have the right to decide the degree of fitness, etc., at which life should be forfeited? Most of us would be glad that it did not stop at the point just above ourselves. We have only one life to live, and we are entitled to live it to its fullest extent. That is true of the unborn fetus. It also has the right to live.

Present society is fast rushing toward that stage that was known in ancient times when a father could decide to destroy his child. One famous papyrus letter from Roman times tells of a father writing an endearing letter to his wife. Their child is soon to be born, and he quite openly says that if it is a girl she is to "expose" it. In other words, it is to be put out to die, thrown away as a piece of rubbish. It is not too much to say that our society is fast rushing toward that state again, with this argument that the mother has the right to decide the fate of her unborn child.

Killing a Baby and Murder

Headlines were made in a recent case where a mother aborted herself with a knitting needle that went through the head of her unborn baby. She was cleared of all charges against her. If the baby had been born and she had then taken similar action, she would have been tried on a criminal charge for murder. No one can say she would have been found guilty, for it is not always possible to predict legal outcomes. In any case, society would have taken a very different approach to the case if her action had been completed after the baby was born, rather than before birth.

Did that woman really have the right to terminate that little life, even excluding the heinous method of death?

We shall see that in many cases of abortion, the fetus is undoubtedly born alive. It is a widely accepted social practice that the medical fraternity has the right to discard those babies so that they will die. Do they have that right of such action? It is not enough to argue that the baby will die anyway, for so would any normal baby if left without attention. So also would a two-year old child.

That Baby is an Invited Guest

Abortion as a general practice can be morally justified only if it be decided that the embryo within the mother's womb is nothing more than a piece of meat in her body. If, however, that embryo is in fact a human life, then all the resources of society should be available to ensure its right to live.

As for the argument that the mother has the right to do what she likes to her own body, this is only true with those cells that are from her own body. When she has ***allowed*** her body to become the home for a new being, she has surrendered her "right" of privacy.

We would recognize that a person has given up her right to privacy if she were abusing or even neglecting a child in her own home. Court orders would soon be issued. When the mother refuses to recognize the right to life of that living being, to whom she has offered her body as a home for a period of several months, then she too has lost her right to "the privacy of her own body." That living being *within* her body is *not her body.* It is a body of a unique human being. Does the mother really have the right to decide to expel her invited guest when that expulsion means the death of that same guest?

Deformed Babies and Abortion

It is relevant to point out that the arguments for abortion do not usually come from organizations of parents where those parents have actually given birth to mentally retarded children. Such parents usually decide to keep the child, and often seek the help of others for the child's welfare. Those organizations seek to find funds and means whereby better educational facilities and help can be given to the retarded. They do not set out to endorse a form of euthanasia by killing children who will be deformed before they actually come to birth. Over and over again, mothers of children with Down's syndrome (mongoloid children) have no thought whatever of disposing of that child, but learn to love it very much, and to live with the situation.

If there was a valid rationale for killing the unborn child because it was believed by tests that it would be defective, then logically the same principles should be extended *after* birth. If we accept such a situation, we have immediately brought ourselves to the place of endorsing euthanasia, the killing of defective people.

Where does it end? Why stop with the young baby? Why not go on through society and get rid of all those who do not measure up to the standards of fitness that are laid down? Once again, who shall decide? Shall it be a board of medical doctors? Or shall we depersonalize it all, and simply let the computer decide?

It would make it much easier, then, if we could develop a computer that could operate the electric chair, or drop the cyanide pellets, or maybe we should teach it to use a guillotine!

Rubella—German Measles

What we are saying applies to even such cases as potential defectives due to rubella (German measles). Where mothers have been infected with rubella during the first 12 weeks of pregnancy, the fact is that less than 20% of the babies will have any defects. Is a decision made to kill the baby because there is at most one chance in five that it would be defective, a justifiable decision? What a responsibility! Someone is actually killing four normal babies unnecessarily, even by this amoral set of values.

If we do accept the principle of killing defective babies, the logical thing is to wait until the babies are born and then kill only those that *actually are* defective! Obviously, we are then back to the same old story—we are endorsing euthanasia when we insist on killing all babies whose mothers have been subjected to rubella (not to mention killing the four out of five that were normal, healthy infants).

What we are saying is that ultimately the concept of abortion and euthanasia are one and the same. If in all seriousness we are to agree to abortion, then we must necessarily as a society endorse euthanasia. Those of us who love our parents and have had the privilege of their living beyond three score years and ten will testify that we wanted them alive beyond that particular time.

As a matter of fact, even that one baby in five that would be disadvantaged because of rubella does not always have problems that are especially serious. About half of such babies with problems are likely to have hearing difficulties that can be corrected by mechanical aids. About half of the disadvantaged children are likely to have heart defects that can be corrected surgically. About 30% will have cataracts, but in the main their vision is likely to be reasonably normal.

About 1% of the normal population suffers from mental retardation, and about 1.5% of the "rubella babies" will be so affected. Clearly, a child should not be aborted because of rubella infections.

A Roman Catholic Argument?

Another aspect relevant to "Who Shall Decide" is that many people wrongly consider the efforts against abortion to be nothing more than a biased religious viewpoint. That being the case, it should be pointed out that the case against abortion is not simply a Roman Catholic argument opposing the rest of the world. The press tends to give that point of view, often linking the so-called Catholic point of view with the moralist or conservative argument. The fact is that Catholics as individuals are indeed concerned, but there are many others as well. The U.S. Supreme Court gave its decision on January 22, 1973, allowing abortion on demand until live birth, and that has caused repercussions which go far beyond any particular group.

A large proportion of the population simply is not aware of the rights of the unborn child, or even of the biological facts as to his existence as a complete human being. If they were so aware, undoubtedly there would be a much louder outcry against abortion: a much stronger voice would be raised for the rights of those innocent little babies.

This is NOT simply deciding for Roman Catholics or some other group. This is a matter of deciding for those little babies, who as yet cannot defend themselves.

Part of the hope of this book is to help raise such an interest, and to increase such an understanding. In November, 1972, the voters of North Dakota, which has only a 27% Catholic population, rejected legalized abortion in a public referendum by 77%. In that same year, Michigan, which is 26% Catholic, voted against legalized abortion by 62%. Despite this, the next year the Supreme Court gave its decision which supported the mother's so-called right to privacy, instead of endorsing the right to life of the unborn child.

The fact is, God sees life as life. The unborn child is a living being in the sight of the Creator God.

Who Then Shall Decide?

Who then shall decide? We, the members of society, should decide. That decision should be made to educate young people in relation to the ethical, moral, and spiritual issues involved with abortion. We should give practical assistance so that economic factors can be faced realistically.

We should decide to provide far more funding for effectively caring for this situation, even if it means reducing the number of fighter planes we provide for Third World countries. It is still true that charity begins at home.

We should decide to be a great moral nation again, adhering to principles whose foundations are established in that Book which was the guide to conduct for our fathers. When that happens, the right to life of every unborn child will be respected.

Idealistic? Yes, and may we be preserved from being a people unable to work toward ideals and goals for the best living.

Who shall decide? You and I. So let us be strong enough to stand up and be counted. The results might surpass our greatest expectations.

Chapter 26

Facts About Abortion are not Pleasant

In this chapter we consider some very unpleasant facts. We touch on the statistics of abortion and the methods of abortion. We shall also consider psychological and other effects on the mother who has had her child aborted.

First we consider relevant statistics about abortions in the United States of America.

An Abortion a Minute in the US!

A pamphlet issued by The Committee of Ten Million, from Glendale, California, in 1973, tells us, "In the nine wars and 198 years since 1775 there have been 667,286 Americans killed in battle. In 1972 alone six hundred thousand babies were killed by abortion—that is more than one murder in a minute for each of the 525,000 minutes in a year. In 1972 alone more babies were killed by 'legal' abortion than American servicemen were killed in the Vietnam, Korean, Spanish, Mexican, 1812, and Revolutionary Wars put together." As the leaflet goes on to say, in war the medical corps attempts to save lives, whereas in abortion the medical profession is brought in to take lives—and what dreadful methods they use!

How They Kill Unwanted Babies

Even physically, abortion is a dreadful thing. There are four methods commonly used. These are:

(a) dilatation and curettage
(b) suction
(c) hysterotomy
(d) saline poisoning

Under (a) the baby is literally cut into pieces by a curette that is inserted into the uterus.

Under (b) a hollow plastic tube is inserted into the uterus, and a powerful suction apparatus is attached to this. Soon the baby is torn into small pieces, and sucked out through the uterus.

Under (c) the baby is removed as in a Caesarian section, with the mother's abdomen opened surgically and the baby lifted out. The baby is alive, and must be put aside and allowed to die if this method is to be "successful." Under other circumstances, many such actions would involve charges of muder. Literally all babies aborted by hysterotomy are alive at birth. They must literally be murdered, or be put aside, soon to die because no attention is given to them.

In the Wilke book, a report is quoted relating to 73,000 abortions. Hysterotomies accounted for 1.3% of the total. The estimate is given that 3,900 babies had been aborted in this manner in New York in 1971, and that all were born alive. They were either allowed to die or were "encouraged" to die.

In method (d) a needle is inserted through the mother's abdominal wall into the amniotic sac of the baby. A concentrated salt solution is then injected. The baby will breathe this fluid and swallow it. Needless to say it dies a death that is horrible, and it takes about one hour as it convulses and struggles. About a day later the mother will give birth to a dead baby.

Dreadful, isn't it? Yes, it is. For instance, the saline method of killing cannot be done before about the sixteenth week of the baby's life, and by then it is indeed a baby in the fullest sense of the work. There are also risks to the mother, with occasional brain damage and lung embolism, which sometimes results in death. The fact that complications sometimes follow abortions is well-known in medical circles.

One grisly side of all this is that if the abortion is by hysterotomy, the baby is born alive. Therefore a full deduction on income tax can be claimed, and is claimed by many, in the United States.

Rape and Abortion

Needless to say, the question of rape is highly relevant in this consideration of abortion. We are not unaware of the problem, but there are other aspects that are often quoted. We again quote the Wilke book: "A scientific study of 3,500 cases of rape, treated in hospitals in the Minneapolis-St. Paul area revealed zero cases of pregnancy. This study took place over a 10-year period."[1]

We are sympathetic to that occasional case where a girl is pregnant because of rape or incest, but to kill the baby is not necessarily the answer. Five points are highly relevant.

1. The baby is innocent, even though the father is guilty of a terrible crime. It certainly is an unfortunate fact for the mother to be raped, but should she have the right to kill an innocent baby because of the monstrous crime of someone who has sinned against her?
2. It is an established fact that far more psycho-

[1]*Op. cit.*, p. 33.

logical problems are likely to follow an abortion than the psychological problems involved in actually giving birth to the baby.

3. If the baby is indeed unwanted when born, he still has the right to live. Adoptive agencies are ready to see that such a child is given a suitable home by parents who desperately want a child.
4. The mother has already gone through her traumatic experience in that unwanted sexual ordeal that was forced upon her. Her main trauma is over, and it is unlikely to help her in the future to know that she was guilty of killing a developing baby inside her own body.
5. If it were to be accepted that abortion could be justified in cases of rape or incest, all that every pregnant girl would need to do to cause her act to be socially accepted would be to insist that she was raped. This is an easy enough claim to make, and who is to prove otherwise? In view of the statistics available, it is a strange fact that an extremely low percentage of those who have been attacked in this way are in fact pregnant by their assailant.

Abortion and Mental Illness

It is a fact that an abortion is more likely to be detrimental in cases of mothers who have mental health problems than nonabortion would be. It is at least arguable that a claim for mental illness is often merely an excuse for abortion. There is seldom proper follow-up in such cases. This is another indication that destroying an unwanted child is often a convenient way of escape for a prospective mother in a difficult social or economic situation.

In addition, after abortion there are undoubtedly psychological effects on a woman that are too seldom reported. The concept of motherhood is inherent in a woman, and the act of abortion violates that inherent sense. When that baby is aborted, in ways that she cannot altogether explain, a basic violation of her womanhood has been effected.

There are far more deaths by suicide after abortion than there are after so-called "unwanted" pregnancies that have been allowed to come to term. Guilt, depression, and self-reproach are common after abortion. These are other aspects that are not given publicity in all that has been written (at great length) about the woman's so-called right to her own body.

It is one thing to remove that struggling little life from the body of its mother, but it is quite another to remove the sense of guilt from the mind of the mother. A large number of women who have had an abortion live with a sense of anguish as a result of the disposal of that living being from within their own bodies. We repeat, there is far more likelihood of psychological disturbance as the result of an abortion than as a result of allowing the child to develop to full term, and then perhaps having it adopted.

Unwanted Pregnancies And Wanted Babies

It is not enough to talk about unwanted pregnancies. A large number of pregnancies within marriage are unwanted, but statistics show that once that baby is born it is just as likely to be wanted as any other child.

As for battered children, the statistics show that this is a fact of life in many families. Furthermore, statistics show that 90% of children who are battered resulted from planned pregnancies and were in no sense "unwanted."

If it be argued that the mother risked her life by giving birth to the child, let it be pointed out that there are even greater risks in abortion. There is the possibility of embolism, and with such methods as dilatation and curettage (whereby the cervical muscle must be paralyzed and stretched before it is ready), obviously there are certain risks. Bleeding from this type of operation is often profuse. If it be argued that there is a risk in giving birth to a child, remember that there is also a risk in aborting that living being.

Abortion and Population Statistics

It cannot be argued that, because of the population explosion, abortion is a legitimate means for birth control. In fact, the birth rate of the United States is decreasing. In 1910 there were approximately 30 children being born per thousand adults, dropping to about 25 per thousand in 1957, and to 17 per thousand in 1970. There have been years of variation, of course, partly because of effects such as "war babies," etc., but the basic facts are that the population is tending now to decrease rather than to increase, so far as the United States is concerned.

In comparison to that, the annual death rate is about 10 per thousand. This means that the population is fast getting to the stage that there are likely to be very many more older people, 65 years and older, in the next generation than there are at that age today. As the population is declining, the average age is increasing.

Abortion and Medical Risks For the Mother

There is a very significant risk to the mother in the case of abortion using the salt perfusion method. This method of abortion has been banned in Japan, but is still practiced in various Western countries. Certainly the risk of death to the mother is much higher by such a method than is the case by the normal process of giving birth to that baby whom she has invited to be a guest in her womb.

When we read that eight out of every ten women aborted need blood transfusions, it is clear that abortion is not just what some glibly refer to as "a little operation." The question might be asked, "Why do we not hear of deaths as a result of transfusions?" Every time a blood transfusion is carried out, there is a risk that hepatitis will be transmitted. Approximately one pint of every 1,000 pints of blood used for transfusion carries that virus, despite all efforts to prevent it. A woman who is hemorrhaging after an abortion is likely to need three or four pints of blood, and one in every 250 women is likely to be infected—and possibly die—of hepatitis. The fact is, her death is not listed as being from an abortion, but from hepatitis.

Of course, only a fraction of hepatitis cases are fatal, but even mild cases lead to lengthy illnesses, and often to some permanent loss of one's stamina and physical well-being.

The process of scraping and cutting, used in dilatation and curretage to remove the embryonic human being, is also likely to cause tiny pieces to be absorbed into the mother's bloodstream. If these travel to such vital areas as her lungs, brain, etc., they can cause serious complications.

Such a process is also likely to affect later pregnancies, for scarring is likely to occur, so that on later occasions a woman's fertilized egg might not be able to attach itself to the wall of the womb. Tubal pregnancies are more likely following such abortions. There is likely to be internal hemorrhaging and emergency surgical procedure, including the removal of that tube. Such tubal pregnancies can take place apart from previous abortions, but the percentage rises from about 0.5% to nearly 4% when there has been an abortion.

Abortion and Sterility

It is also established that a woman who has had an abortion is very much more likely to have a spontaneous miscarriage with later pregnancies. About 10% of those who have had abortions find they are sterile, often a consequence that was not planned when that first abortion was performed. Even where a pregnancy does develop, it is three times more likely that the baby will be born prematurely when the mother has had an abortion previously. When an abortion takes place, muscle fibers are torn, and undoubtedly a weakening takes place.

Not only the mother who has had an abortion pays a price, so also does the hopeful father, especially where he had nothing to do with the earlier disposal of an unwanted human being. He now pays a price that should not be asked of him. To a greater or lesser degree (varying with individual cases), parents and other relatives also suffer emotional and other psychological consequences.

The facts about abortion are not pleasant. They are heartrending.

Chapter 27

"But it's so small!"

Some people seem to think that the embryo is not a human being, simply because it is "so small." However, there is no point where it can be shown that the fetus is a baby "now," but was *not* a day or a week before.

In their outstanding book, *Handbook on Abortion* (referred to previously), Dr. and Mrs. J. C. Wilke quote from the report of the First International Conference on Abortion, held in Washington, D.C., in October, 1967. Authorities came together from such diverse fields as medicine, law, ethics, and social sciences. Within the medical group there were biochemists, professors of obstetrics and gynecology, geneticists, and others. Various races and religions were brought together, with only 20% being Roman Catholic. Dr. and Mrs. Wilke quote from the report at one point as follows: "The changes occurring between implantation, a six weeks' embryo, a six months' fetus, and a one-week old child, or a mature adult, are merely stages of development and maturation" (p. 8).

The Genetic Plan is There at Conception

The fact is that the male sperm contributes 50% of that new life within the mother's womb, and the female ovum contributes the other 50%. The sperm carries the genetic information from the father, while the ovum does the same for the mother. When the two come together, 23 chromosomes from the father join with 23 chromosomes from the mother, and the being so created is an entirely new life, *not a part of the mother's body*. That new being is unique, different from every other being that has ever lived in the universe, or ever will so live.

This is true even with identical twins. Actually, when the cells divide in the way that identical twins are produced, the second twin is the offspring of that first cell.

It is interesting to note that an identical twin is a division from an original cell. That division must occur between the time when the ovum is fertilized and before it is implanted in the wall of the mother's uterus. Such a division will never take place after implantation. From that point of development, in the mysterious formation of life, each twin is a unique being.

It is not enough to say that the beginning fetus does not look like a human. As a matter of fact, the electron microscopes that are available today go far beyond what we would see with normal eyesight, and a miniature though immature person is there. The whole life that will one day be a mature adult is totally contained within that tiny embryonic life package.

If despite all the arguments available to indicate that the fetus is in the fullest sense a human life, there is still doubt in the minds of some, such doubt does not justify abortion.

That Baby Fights for Life

Some of us have seen desperate attempts to keep somebody alive. A dying young man, a child rescued from the sea, a baby that has had convulsions, and many others. The fact is that we recognize the right of each of those people to remain alive. We give them the benefit of the doubt, as it were, even when sometimes it might be thought that they are not especially useful members of society.

At least that "same benefit of the doubt" should be given to a new life that is obviously unable to fight successfully when the adults by whom it is surrounded (perhaps even one adult—his mother) decide to kill it.

It certainly is true that this baby fights as much as he can. We have already said that he has developed extra parts for his body to cope with the requirements of life within his mother's body. That umbilical cord is a lifeline, the amniotic sac is his own special space capsule, and the placenta is his life-support system. It is the baby who decides when to be born, but sadly it is often the mother or other advisers who decide when his life will be terminated.

The baby will have a heartbeat between the 18th and 25th day. Electrical brain waves have been recorded as early as 43 days, with the brain itself complete in miniature form by about 8 weeks.

"A Natural Swimmer's Stroke"—At Two Months

In the Wilke book there is a fantastic story derived from the writing of Paul E. Rockwell, M.D., Director of Anesthesiology at the Leonard Hospital in Troy, New York. He tells of giving an anesthetic in the case of a ruptured tubal pregnancy at two months. The doctor was handed what he believed was the smallest human being that human eyes had ever seen. The embryo sac was still intact, and was transparent. Within that sac was a tiny human male, only one third of an inch in length, but "swimming extremely vigorously in the amniotic fluid, while attached to the wall by the umbilical cord. This tiny human was perfectly developed with long, tapering fingers, feet, and toes

. . . . The baby was extremely alive, and swam about the sac approximately one time per second with a natural swimmer's stroke. This tiny human did not look at all like the photos and drawings of embryos which I have seen, nor did it look like the few embryos I have been able to observe since then, obviously because this one was alive. When the sac was opened, the tiny human immediately lost his life and took on the appearance of what is accepted as the appearance of an embryo at this stage (blunt extremities, etc.)" (p. 18).

That "tiny human" is just that—a human being. If one of our own body cells dies, that is part of us dying, but when a fertilized ovum dies, that is an entire person who dies. That tiny embryonic human being has fantastic capacities, as we have already seen, to produce its own life support systems. It even has the power to suppress its own mother's menstrual periods.

Another amazing fact is that there is a relationship between the mother and the baby that makes it possible for them to tolerate each other for nine months, despite all the problems of immunology. After that baby is born, he could not necessarily accept a skin graft or blood from that same woman who bore him for nine months, and yet for that period of time she accepts him and he accepts her. There is no suggestion of rejection due to one being foreign to the other.

The Embryo is a Child in God's Sight

There is another very strong argument to show that the embryo is fully hunan: it is so in God's sight. That is made clear in our next chapter where we consider a number of relevant passages in the Bible.

Chapter 28

What the Bible Teaches

Throughout this book we have referred to the Bible a number of times. That Book is recognized by Christians as the final authority in matters of faith and conduct.

In this chapter we shall list a number of verses, quoting from the New American Standard Version. We shall give brief comments to show the relevance to the subject matter we have discussed in this book. The list could be expanded considerably.

Man in the Image of God

> **Genesis 1:27:**
> And God created man in His own image, in the image of God He created him; male and female He created them.

Man was made in the image of God. This is not said of any lesser creature. Some aspects of genetic engineering offend this divinely established principle. This is especially true of experiments where genetic structures of animals and men are mixed.

> **Genesis 2:23:**
> And the man said, "This is now bone of my bone, and flesh of my flesh; she shall be called woman, because she was taken out of man."

No better explanation of woman's origin has ever been given. She is one with man, yet distinct from man.

God Has Set Certain Limits

Genesis 3:24:

So He drove the man out; and at the east of the garden of Eden He stationed the cherubim, and the flaming sword which turned every direction, to guard the way to the tree of life.

Man was not allowed to partake of the Tree of Life. It was for man's good that he was refused access to that source of life. Sin had intruded, and the original life was now tainted. God had a better plan, with the eternal life to be offered by Jesus Christ. In a coming age men will indeed partake of that tree (Revelation 2:7; 22:2).

Genesis 11:6, 7:

And the Lord said, "Behold, they are one people, and they all have the same language. And this is what they began to do, and now nothing which they purpose to do will be impossible for them.

"Come, let Us go down and there confuse their language, that they may not understand one another's speech."

Men were going beyond the bounds set by God, and judgment fell. If cloning also goes beyond the set limits, God's judgment will again fall, in God's own time. The Tree of Life is still "forbidden fruit."

The Taking of Life is Murder

Genesis 9:6:

Whoever sheds man's blood, by man his blood shall be shed, for in the image of God

He made man.

The taking of human life by an individual is a heinous offense before God. (See also Psalm 94:20, 21; Matthew 5:21, 11; I Peter 4:15; I John 3:15; Revelation 2:8; 22:15.)

The Bible especially mentions the killing of the innocent (Exodus 23:7). One who performs such an act is there declared "Guilty." In Proverbs 6:16, 17 we learn that the shedding of innocent blood is especially mentioned as an abomination before God.

Another relevant passage is Deuteronomy 27:24 where a curse is declared against one who strikes his neighbor in secret. What closer "neighbor" could there be than that guest in the mother's womb? At Matthew 22:39 Jesus reminds us that we are to love our neighbor as ourselves.

Would we kill ourselves? Could THAT be justified in God's sight? Of course not—and neither can abortion.

What About the Handicapped?

Psalm 51:5:

> Behold, I was brought forth in iniquity,
> and in sin my mother conceived me.

Other Scriptures make it clear that the marriage relationship is honorable "and the bed undefiled" (Hebrews 13:4). Therefore this verse from the Psalms cannot be a condemnation of the sex act in honorable marriage. The fact is, we are members of a fallen race, and every single human being has defective genes. That is a consequence of "the fall."

It stands to reason that some will be more affected than others—hence there will be "disadvantaged" children.

Exodus 4:11:

> And the Lord said to him, "Who has
> made man's mouth? Or who makes him

> dumb or deaf, or seeing or blind? Is it not I, the Lord?"

The Scriptures proclaim God as the "ultimate (or Final) Cause." Hence He is the One Who allows some to be dumb, others deaf, others again blind.

> **I Samuel 16:7:**
>
> But the Lord said to Samuel, "Do not look at his appearance or at the height of his stature, because I have rejected him; for God sees not as man sees, for man looks at the outward appearance, but the Lord looks at the heart."

Saul was handsome and attractive with his fine physique. Others less endowed, however, would sometimes be preferable to God. He looks beyond the physique to what a man is in his spiritual attitudes. At Matthew 18:8 we read that the Lord says, "It is better for a man to enter life crippled or lame, than having two hands or two feet, to be cast into the eternal fire."

This certainly is not an endorsement of abortion or of abandonment of a disadvantaged child! That is indirectly indicated in our next verse. Jesus said:

> **Luke 14:13, 14:**
>
> But when you give a reception, invite the poor, the crippled, the lame, the blind, and you will be blessed, since they do not have the means to repay you; for you will be repaid at the resurrection of the righteous.

Notice that there is blessing for those who are considerate to the crippled, the lame, and the blind. Jesus would not have endorsed euthanasia or abortion. Paul tells us that widows over 60 years of age are to be especially helped by the Church (I Timothy 5:9). Such consideration for the elderly is directly opposed to the concept of euthanasia.

What of Those Who Have "Disadvantaged" Children?

The Bible makes it clear that the one with children is blessed of the Lord. We read:

Psalm 127:3-5:

> Behold, children are a gift of the LORD; The fruit of the womb is a reward. Like arrows in the hand of a warrior, so are the children of one's youth. How blessed is the man whose quiver is full of them; they shall not be ashamed, when they speak with their enemies in the gate.

At no point is there any indication that deformed children are in a different category. For all children there are positive statements.

Job 10:8-12:

> Thy hands fashioned and made me altogether, and wouldst Thou destroy me? Remember now, that Thou hast made me as clay; and wouldst Thou turn me into dust again? Didst Thou not pour me out like milk; and curdle me like cheese; clothe me with skin and flesh, and knit me together with bones and sinews? Thou hast granted me life and lovingkindness; and Thy care has preserved my spirit.

That little baby was formed by God, clothed with skin and flesh, knit together with bones and sinew. Here a similar picture is given.

Psalm 139:13-16:

> For Thou didst form my inward parts; Thou didst weave me in my mother's womb. I will give thanks to Thee, for I am fearfully and wonderfully made; wonderful are Thy works, and my soul knows it very well. My frame was not hidden from Thee, when I was made in secret, and skillfully wrought

> in the depths of the earth. Thine eyes have seen my unformed substance; and in Thy book they were all written, the days that were ordained for me, when as yet there was not one of them.

That babe was "skillfully wrought," his "inward parts" woven by God in the mother's womb.

What then of the mother and the father who have a disadvantaged child as their offspring? Undoubtedly they are tested, but God has promised the Christian that he will not be tempted beyond what he can bear, but will provide a way of escape (I Corinthians 10:13).

The Apostle Paul himself had a "thorn in the flesh," and we are not told what it was. He asked the Lord three times to remove it, but was told that God's grace was sufficient—that His strength was perfected in weakness (II Corinthians 12:7-9). God will give "grace to help in time of need" (Hebrews 4:16).

When There is an Induced Miscarriage

Here is another Bible passage that needs comment.

Exodus 21:22-25:

> And if men struggle with each other and strike a woman with child so that she has a miscarriage, yet there is no further injury, he shall surely be fined as the woman's husband may demand of him; and he shall pay as the judges decide.
>
> But if there is any further injury, then you shall appoint as a penalty life for life, eye for eye, tooth for tooth, hand for hand, foot for foot, burn for burn, wound for wound, bruise for bruise.

This is NOT saying that there is to be no penalty if there is a miscarriage whereby a child is lost. As the marginal reading of verse 22 informs us, the literal reading is "so that her children come out"—they are born prematurely, but live. This is when there is NO loss of a child. The passage actually makes it clear that when there IS the loss of a child, then the one responsible is just as guilty as though he had killed an adult. The penalty is "life for life" (verse 23).

We mention in passing that the Code of Moses was immeasurably superior to that of Hammurabi. With Hammurabi, the penalties varied according to social status. With Moses it was equality—"an eye for an eye, and a tooth for a tooth." No vendettas could be permitted, no feuds whereby ten lives might be demanded for one. "A tooth for a tooth," the so-called *lex talionis,* was not primitive or barbarous. It was a pointer to the equal standards of justice that God demanded.

The final passage on which we wish to comment is very important.

Matthew 1:18:

> Now the birth of Jesus Christ was as follows. When His mother Mary had been betrothed to Joseph, before they came together she was found with child of the Holy Spirit.

Are we exaggerating when we talk about God seeing the human embryo as fully human? At Matthew 1:8 to 20 we read of the birth of Jesus. We saw at verse 18 that Mary "was found to be with child by the Holy Spirit." Remember, that was before the birth of our Lord and Savior—the Bible describes Him as a "Child."

At Luke 1:41 we read of John the Baptist: "When Elizabeth heard the salutation of Mary, the babe leaped in her womb." This is not just described as a fetus, but as "the babe." That is the same Greek word that is used in Luke 2:12-16 to describe Christ in a manger. In fact, the same word is translated "babe" in I Peter 2:2, "child" in II Timothy 3:15, "infant" in Luke 8:15, and finally "young child" in Acts 7:19.

The Scriptures make no distinction between the unborn child and the child after birth, as to it being a child. At Jeremiah 1:4, 5, we read that God knew Jeremiah before he was formed in the belly, and that he was set apart before he came out of the womb. Isaiah was formed from the womb to be God's servant (Isaiah 49:5). See also Job 10:11, Galatians 1:15, and Psalm 139:13-16 where similar statements are recorded about Job, Paul, and David.

We repeat. In God's sight that embryo is a living being, described by the same word as an elder child, long since born.

Summary

The authors of this book see the potential blessings of genetic engineering, but we also recognize the serious implications of recombinant DNA.

We agree that "test-tube babies" can bring happiness to parents where a mother has blocked fallopian tubes. However, we also believe that man does not have the right of choice as to which embryo to accept and which to destroy, for they are all complete life forms as soon as the ovum and the sperm have united.

We oppose abortion in every circumstance where life could have been developed to the point of the child being born. Obviously, this would not be the case with a tubal pregnancy where such development is out of the question.

We oppose all forms of deliberate destruction of newly-born babies, even where there are physical or mental abnormalities.

We recognize that this last statement could involve sacrificial self-giving by some parents. One of your authors (Wilson) has seen that as a result of an accident to one of his own brothers. The sacrificial self-giving of his parents for 13 years was an experience whereby many would later rise to call them blessed.

We believe that Christians should be informed in these areas and be prepared to present their convictions to others.